Integrating Results

through Meta-Analytic Review
Using
SAS® Software

Morgan C. Wang & Brad J. Bushman

The correct bibliographic citation for this manual is as follows: Wang, Morgan C. and Bushman, Brad J. 1999. *Integrating Results through Meta-Analytic Review Using SAS® Software.* Cary, NC: SAS Institute Inc.

Integrating Results through Meta-Analytic Review Using SAS® Software

Table of Contents

Using This Book

Who Should Use (Buy!) This Book

This book was written for the practicing meta-analyst. It was written to show individuals *how* to carry out the data analysis portion of a meta-analysis from beginning to end (without becoming frustrated and confused in the process, we hope). We assume that the reader has an understanding of basic statistical methods (for example, analysis of variance, regression analysis). Readers who lack a basic understanding of statistical methods can consult some good reference books (for example, Agresti, 1990; Chambers, et al., 1983; Neter, Wasserman, & Kutner, 1990; Snedecor & Cochran, 1989). We also assume that the reader has a basic understanding of meta-analytic procedures. Readers who lack a basic understanding of meta-analytic procedures can consult some good reference books (for example, Cooper & Hedges, 1994; Hedges & Olkin, 1985; Hunter & Schmidt, 1990). A basic understanding of SAS software is also desirable. Chapter 2 of this book describes the SAS procedures that you can use to conduct a meta-analysis and contains additional references for SAS software.

Organization of the Book

The SAS software in this book is much more flexible and powerful than any existing software for conducting meta-analytic reviews. It is also quite user friendly. This book draws heavily from the *Handbook of Research Synthesis* (Cooper & Hedges, 1994), the state-of-the-art meta-analysis reference source (see Glass, 1995, for a review). This book contains the SAS code to carry out most of the meta-analytic procedures that are described in the *Handbook*. But this book is not just a SAS supplement to the *Handbook*. It takes a more "hands-on" approach than does the *Handbook*. It also contains important material that is not included in the *Handbook*, such as a chapter on how to combine effect-size estimates and vote-counts (Chapter 7), a chapter section on random-effects ANOVA models (Chapter 9, Section 9.2), a chapter section on using normal quantile plots to explore meta-analytic data sets (Chapter 3, Section 3.4), and a chapter section on controlling for the effects of covariates when combining odds ratios (Chapter 4, Section 4.7).

We believe that examples are important when introducing technical material. This book includes lots of examples to help you understand important concepts in meta-analysis. Although a few of the examples use simulated or hypothetical data, most of the examples come from the empirical literature in the social sciences, education, and medicine. We encourage you to try the examples in the book before using the meta-analytic procedures on your own data set. You can use SAS Online Samples to access the data sets for the examples that are included in this book. See the section "How to Use SAS Online Samples" for more information.

The organization of chapters in the book is as follows:

Chapter 1   is the introduction.

Chapter 2   describes the SAS procedures that are used to conduct a meta-

Chapter 3   analysis describes how to present the results of a meta-analysis in graphical form so that the audience can readily comprehend the most important findings.

Chapters 4   focus on inferential procedures in meta-analysis.
through 10

Chapters 4   describe how to combine effect-size estimates based on categorical
and 5   and continuous data, respectively.

Chapter 6   describes vote-counting procedures, which are used when some research reports do not include enough information to calculate an effect-size estimate but do include information about the direction and/or statistical significance of results.

Chapter 7	describes a recent procedure for dealing with missing effect-size estimates. This procedure, which combines effect-size estimates and vote-counts, is more effective in handling missing effect-size estimates than other procedures that have been proposed (for example, discarding studies with missing effect-size estimates, setting missing effect-size estimates equal to zero).
Chapters 8 and 9	deal with fixed- and random-effects models, respectively.
Chapters 4 through 9	assume that the effect-size estimates to be combined are independent.
Chapter 10	describes how to combine dependent or correlated effect-size estimates using multivariate procedures.
Chapter 11	describes how to report the results from a meta-analysis, and it gives an example of how to conduct the data analysis portion of a meta-analysis.

Good luck as you begin the adventure of conducting a meta-analytic review. We hope this book serves as a useful guide on your adventure.

Conventions

In this book, the following terms are used to discuss SAS syntax: (SAS, 1990, pp. xxi–xxiii):

keyword	is a literal that is part of the SAS language. A literal must be spelled exactly as shown, in either uppercase or lowercase letters (SAS language is not case sensitive). Keywords in this book are SAS statement names.
argument	is a literal or user-supplied element that follows a keyword. Some arguments are required, whereas other arguments are optional. Optional arguments are enclosed in angle brackets like this: *<argument>*.

value   is a literal or user-supplied element that follows an equal sign. Values
                   are assigned to arguments.

Style Conventions

We use the following style conventions to explain SAS syntax:

UPPERCASE BOLD   identifies SAS keywords such as the names of
                          statements (for example, **CLASS**) and procedures
                          (for example, **PROC PRINT**).

UPPERCASE REGULAR   identifies arguments and values that are literals (for
                          example, PLOT).

lowercase italic   identifies user-supplied arguments and values.

These style conventions are only used to explain SAS syntax sections. Elsewhere
in the text, we use UPPERCASE REGULAR font to identify SAS keywords,
arguments, and values.

All SAS variable names begin with a letter. All SAS statements end with a
semicolon.

How to Use the SAS Code in This Book

Each chapter in this book has two directories, one for the SAS code and another for
the SAS data sets that are used in the chapter. In Chapter 2, for example, the
directory for the SAS code is called D:\METABOOK\CH2\SASCODE, and the
directory for the SAS data sets is called D:\METABOOK\CH2\DATASET.

For each chapter in this book (except Chapters 1 and 3), we use the following
convention for naming SAS code for examples: EX-*chapter-number-example-
number*.SAS. For instance, the SAS code for Example 2.4 is called EX24.SAS.
This example would be in the SAS code directory D:\METABOOK\CH2\
SASCODE. Similarly, we use the following convention for naming two-level (that
is, permanent) SAS data sets for examples: CH-*chapter-number*.EX-*chapter-
number-example-number*. For example, the SAS data set for Example 2.7 is called

CH2.EX27. This SAS data set would be in the SAS data library D:\METABOOK\CH2\DATASET.

Chapter 1 contains no computational examples. In Chapter 3, we use similar conventions, but instead of EX we use FIG and OUTP for figures and output, respectively.

SAS Sample Library

One nice feature of using SAS in a Windows environment is that it comes with the SAS Sample Library. The SAS Sample Library contains an extensive set of sample programs. You can copy the sample programs to the PROGRAM EDITOR window, modify them, and save them. The following steps are required to access the SAS Sample Library (Gilmore, 1996b, pp. 36–40):

1. Press the "Help" button on the SAS System main menu.
2. Click on "SAMPLE PROGRAMS."
3. Double-click on "SAS Sample Library" in the Help Topic: SAS Sample Library window.
4. Click on the topic of interest in the SAS Sample Library window. Each topic contains sample programs.

You can use the following steps to copy a sample program to the PROGRAM EDITOR window:

1. Select the "Copy" option from the "Edit" menu in the SAS Sample Library window's menu.
2. Select the "Close" option under the "File" menu to close the SAS Sample Library.
3. Click on the PROGRAM EDITOR window to activate it.
4. Press the "Paste" button on the SAS System "Edit" menu. The program should appear in the PROGRAM EDITOR window.
5. Modify the sample program as needed.

To save the sample program, press the "Save a SAS Program" button on the SAS System main menu.

How to Use SAS Online Samples

The SAS code for each example that is used in this book is available through the SAS online sample library. The following steps are required to access SAS Online Samples:

1. Use a web browser (for example, Netscape Navigator, Internet Explorer) to access the SAS home page using the following uniform resource locator (URL): /http://www.sas.com/.
2. Select the "Publications" option from the SAS home page.
3. Select the "SAS Online Samples" option from the "Online Documentation" list.
4. Select this book, "Integrating Results through Meta-Analytic Review Using SAS Software," from the list of books with sample code.
5. Select the desired chapter number and example number.

You can then copy the SAS code, paste it in the PROGRAM EDITOR window, save it, and submit it.

Hardware Requirements

In this book, we assume that you are running Windows 97, Windows 95, or Windows 3.1 on a PC. The SAS code in this book, however, will run in any environment.

References

Agresti, A. (1990). *Categorical data analysis*. New York: Wiley.

Chambers, J. M., Cleveland, W. S., Kleiner, B., & Tukey, P. A. (1983) *Graphical methods for data analysis*. Belmont, CA: Duxbury Press.

Cooper, H. M. & Hedges, L. V. (Eds.). (1994). *The handbook of research synthesis*. New York: Russell Sage Foundation.

Glass, G. V. (1995). The next-to-last word on meta-analysis. (Review of the book *The Handbook of Research Synthesis*.) *Contemporary Psychology, 40*, 736–738.

Hedges, L. V. & Olkin, I. (1985). *Statistical methods for meta-analysis*. New York: Academic Press.

Hunter, J. E., & Schmidt, F. L. (1990). *Methods of meta-analysis: Correcting error and bias in research findings*. Newbury Park, CA: Sage.

Neter, J., Wasserman, W., & Kutner, M. H. (1990). *Applied linear statistical models* (3rd ed.). Homewood, IL: Irwin.

SAS Institute Inc. (1990). SAS procedures guide, version 6, 3rd ed. Cary, NC: Author.

Snedecor, G. W. & Cochran, W. G. (1989) Statistical methods (8th ed.). Ames, IA: Iowa State University Press.

chapter 1

Introduction

1.1 Narrative (Qualitative) and Meta-Analytic (Quantitative) Literature Reviews

Science is built up with fact, as a house is with stone. But a collection of fact is no more a science than a heap of stones is a house.

— Jules Henri Poincare (cited in Olkin, 1990)

. . . it is necessary, while formulating the problems of which in our advance we are able to find the solutions, to call into council the views of those of our predecessors who have declared an opinion on the subject, in order that we may profit by whatever is sound in their suggestions and avoid their errors.

— Aristotle, *De Anima*, Book 1, Chapter 2
(cited in Cooper & Hedges, 1994)

All scientists acknowledge that their efforts should build upon past work through replication, integration, extension, or reconceptualization. It is, therefore, ironic that the traditional review of scientific data has typically been conducted in an unscientific fashion. In the traditional narrative (qualitative) review, the reviewer uses "mental algebra" to combine the findings from a collection of studies and describes the results verbally. Statisticians were the first scientists to advocate alternative methods for combining research findings. These methods were labeled *meta-analysis* by Gene Glass (1976):

Meta-analysis refers to the analysis of analyses . . . the statistical analysis of a large collection of analysis results from individual studies for the purpose of integrating findings. It connotes a rigorous alternative to the casual, narrative discussions of research studies which typify our attempts to make sense of the rapidly expanding literature (p. 3).

The quantification of research evidence is the key factor that distinguishes a meta-analytic review from a narrative review (Olkin, 1990). In the meta-analytic review, the meta-analyst uses statistical procedures to integrate the findings from a collection of studies and describes the results using numerical effect-size estimates.

One weakness of narrative reviews is that they may be more susceptible to the subjective judgments, preferences, and biases of a particular reviewer's perspective than meta-analytic reviews. As Glass (1976) states:

> *A common method for integrating several studies with inconsistent findings is to carp on the design or analysis deficiencies of all but a few studies - those remaining frequently being one's own work or that of one's students or friends - and then advance the one or two "acceptable" studies as the truth of the matter (p. 4)*

It is worth noting that inconsistent findings in a meta-analytic review are not necessarily problematic. Inconsistent findings may simply reflect opposite tails of the same distribution of effects. Consider, for example, the following distribution of standardized effect-size estimates (that is, effect-size estimate divided by its corresponding estimated standard deviation) that is centered at 0.50 (Cohen's, 1988, conventional value for a medium-sized effect). By random chance some studies (about 31%) should have negative effects even if the true effect-size in the population is 0.50.

Figure 1.1 Distribution of standardized effect-size estimates centered at 0.50

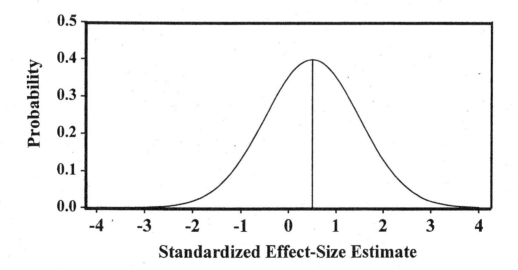

Alternatively, inconsistent findings may imply that some variable moderates the treatment effect. A moderator variable influences the strength and/or direction of the relation between the independent variable (that is, the treatment) and the dependent variable (that is, the response; see Baron and Kenny, 1986, for a discussion of moderator variables). In Figure 1.2, the treatment has a negative effect on Group 1 and a positive effect on Group 2. In this example, most negative effects would be found for Group 1, and most positive effects would be found for Group 2. If group is ignored, however, you might conclude that the findings are inconsistent and that the treatment has no effect.

Figure 1.2 *Distribution of standardized effect-size estimates for two different groups that are affected in opposite ways by the treatment*

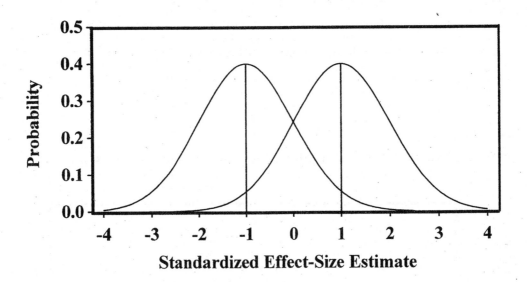

A second weakness of narrative reviews is that they often ignore the magnitude of the treatment effect. In a narrative review, the reviewer frequently uses *p*-values to draw conclusions by counting the number of studies that found significant treatment effects. But *p*-values cannot be used to determine the magnitude of a treatment effect. Consider the following example in which a treatment group is compared to a control group. Assume that the experimental and control groups have equal sample sizes. Which treatment effect is largest: (a), (b), or (c)?

(a) $t(256) = 4.0, p < .0001$

(b) $t(64) = 2.0, p < .05$

(c) $t(4) = 0.5, p < .64$

You may be tempted to answer (a) because it has a smaller p-value, but this is actually a "trick question." It turns out that the treatment effects are identical for all three tests — the effect-size estimate is 0.50 in each case. A formula for obtaining an effect-size estimate from an independent sample t-test is

$$d = \frac{2t}{\sqrt{df}} \tag{1.1}$$

where d is the effect-size estimate and df are the degrees of freedom (Friedman, 1968). Plugging the values for options (a), (b), and (c) into Equation 1.1, you obtain:

$$d = \frac{2(4.0)}{\sqrt{256}} = \frac{2(2.0)}{\sqrt{64}} = \frac{2(0.5)}{\sqrt{4}} = 0.50.$$

The point is that p-values cannot be used as surrogate effect-size estimates.

These weaknesses of narrative reviews can cause their conclusions to be inconsistent with the data. In a study by Cooper and Rosenthal (1980), faculty members and upper-level graduate students in psychology were randomly assigned to use narrative or statistical procedures to review seven studies on sex differences in persistence. None of the reviewers were familiar with meta-analytic techniques. Participants in the statistical group were instructed how to combine the results from the studies. Participants in the narrative group were asked to "employ whatever criteria you would use if this exercise were being undertaken for a class term paper or a manuscript for publication." Participants were asked whether the evidence supported the conclusion that females were more persistent on tasks than males were. Five possible responses were provided (*definitely yes, probably yes, impossible to say, probably no,* and *definitely no).* The results showed that 68% of the statistical reviewers were at least considering rejecting the null hypothesis, compared with only 27% of the traditional reviewers. (The null hypothesis should have been rejected at the .05 level because the confidence interval for the effect size excluded the value zero.) Participants also were asked to estimate the magnitude of sex differences in persistence. Six possible responses were provided (*very large, large, moderate, small, very small,* and *none at all*). The results showed that 58% of the statistical reviewers estimated at least a small sex difference in persistence, compared with only 27% of the traditional reviewers. (The effect was about equal to Cohen's, 1988, conventional value for a "small" effect.) Thus, participants in the narrative group underestimated the presence and the strength of sex differences in persistence.

In the world outside of the controlled laboratory setting, similar results have been reported. For example, an article in *Science* (Mann, 1994) compares the conclusions drawn from meta-analytic versus traditional literature reviews in five subject areas: (a) psychotherapy, (b) delinquency prevention, (c) school funding, (d) job training, and (e) reducing anxiety in surgical patients. The comparison reveals that narrative reviews underestimate the presence and the strength of treatment effects for each subject area. More recently, Hunt (1997) provided several examples of how narrative reviews underestimate the presence and magnitude of treatment effects. Because of their superiority over narrative reviews, it appears that meta-analytic reviews are here to stay. The next section documents the increasing use of meta-analysis.

1.2 Increasing Use of Meta-Analysis

The use of meta-analysis has increased dramatically in recent years, especially in the social sciences, medicine, and education. For example, we tabulated the number of journal articles in *PsycLit* (a psychological research database) and

Medline (a medical research database) that used the keyword "meta-analysis" from 1976 (the year the term was introduced) to 1995. In both databases, the number of entries has increased rapidly and consistently (see Figures 1.3 and 1.4).

Figure 1.3 Increase in the use of meta-analysis over time in psychology

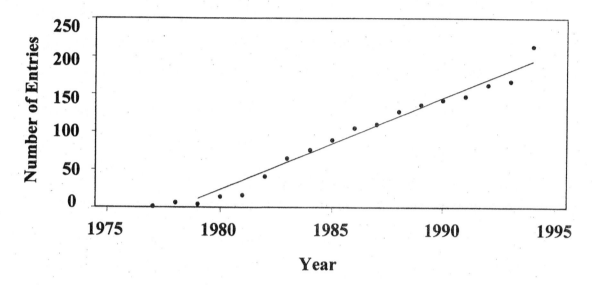

Figure 1.4 Increase in the use of meta-analysis over time in medicine

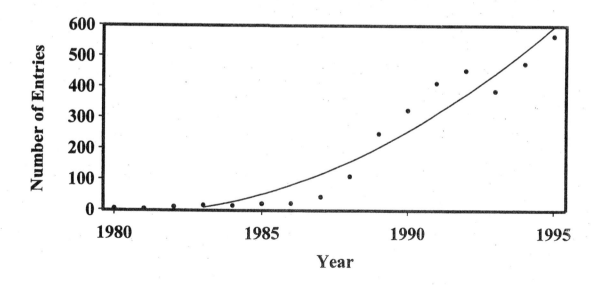

The rapid increase in the use of meta-analysis is likely to continue. In his review of meta-analytic methods, Bangert-Drowns (1986) states the following:

Meta-analysis is not a fad. It is rooted in the fundamental values of the scientific enterprise: replicability, quantification, causal and correlational analysis. Valuable information is needlessly scattered in individual studies. The ability of social scientists to deliver generalizable answers to basic questions of policy is too serious a concern to allow us to treat research integration lightly. The potential benefits of meta-analysis method seem enormous. (p. 398)

1.3 Two Approaches to Conducting a Meta-Analysis

Although the term meta-analysis was coined relatively recently, statisticians have been using these methods for about 100 years. Two different statistical approaches have been used to combine evidence from primary studies. One approach relies on testing the statistical significance of combined results across studies, and the other approach relies on estimating the magnitude of combined results across studies. Fisher (1932), Pearson (1933), and Tippett (1931) were among the first to propose methods for testing the statistical significance of combined results across studies. Consider, for example, the following quotation from the fourth edition of Sir R. A. Fisher's influential text *Statistical Methods for Research Workers*:

When a number of quite independent tests of significance have been made, it sometimes happens that although a few or none can be claimed individually as significant, yet the aggregate gives an impression that the probabilities are on the whole lower than would often have been obtained by chance. It is sometimes desired, taking account only of these probabilities, and not of the detailed composition of the data from which they were derived, which may be of very different kinds, to obtain a single test of the significance of the aggregate, based on the product of the probabilities individually observed (p. 99).

An early application of this approach was described by Stouffer and his colleagues (Stouffer, Suchman, DeVinney, Star, & Williams, 1949). In three studies, male soldiers rated how much they wanted their sisters to join the United States Army. The ratings were used to determine male soldiers' attitudes toward female soldiers. Some of the male soldiers had female soldiers in their own camp, and some did not. In all three studies, male soldiers were less likely to want their sisters to join the Army when there were female soldiers at their own camp. Stouffer and his colleagues combined the p-values from the three studies to obtain an overall significance test.

Significance tests of combined results are sometimes called omnibus or nonparametric tests because they do not depend on the distribution of data. Omnibus tests depend only on the fact that the p-values are uniformly distributed between the values 0 and 1.00 when the null hypothesis is true and the treatment has no effect (see Hedges & Olkin, 1985, p. 2). The primary disadvantage of omnibus tests is that they cannot provide estimates of the magnitude of treatment effects across studies.

Birge (1932), Cochran and Yates (Cochran, 1937, 1943; Yates & Cochran, 1938), and Pearson (1904) were among the first to propose methods for estimating the magnitude of treatment effects across studies. For example, Karl Pearson (1904), the famous biometrician, conducted an empirical review of 11 studies that had tested the effectiveness of a typhoid vaccine. Five studies tested whether the vaccine reduced the incidence of typhoid, and the other six studies tested whether the vaccine reduced mortality among those who had contracted typhoid. Pearson computed average correlations of .23 and .19 for typhoid incidence and mortality, respectively. Pearson concluded that these average correlations were too low to warrant adopting the vaccine for British soldiers: "I think the right conclusion to draw would be not that it was desirable to inoculate the whole army, but that improvement in the serum and method of dosing, with a view to a far higher correlation, should be attempted" (p. 1245).

For at least 50 years, social and statistical scientists have questioned the utility of significance testing in research (for example, Bakan, 1966; Berkson, 1938; Carver, 1978; Cohen, 1994; Falk, 1986; Harris, 1991; Hogben, 1957; Kirk, 1996;

Kupfersmid, 1988; Meehl, 1978; Morrison & Henkel, 1970; Nunnally, 1960; Schmidt, 1996; Shaver, 1993). Even Frank Yates (1951), a colleague and friend of R. A. Fisher, said that Fisher's text *Statistical Methods for Research Workers* "has caused scientific research workers to pay undue attention to the results of significance tests . . . and too little (attention) to the estimates of the magnitude of the effects they are estimating" (p. 32). A common theme emerges from these writings: People often use p-values as surrogate effect-size estimates (for example, they incorrectly assume that small p-values denote large treatment effects). People often misinterpret a p-value as the probability that the null hypothesis is false.

Notwithstanding the attacks social and statistical scientists have waged on significance testing, many people continue to "worship" p-values (Schulman, Kupst, & Suran, 1976). In a humorous article, Salsburg (1985) concluded that far too many physicians are adherents of a religion called Statistics. According to Salsburg, adherents of this religion engage in the ritual known as "hunting for p-values." If the p-value is larger than .05, the practitioner must be prepared to suffer the wrath of the angry gods of Statistics. The deep mysterious symbols of this religion are *ns*, * ($p < .05$), ** ($p < .01$) and (mirabile dictu) *** ($p < .001$). The more *'s, the happier are the gods of Statistics. We think that it is a bad idea to worship p-values because any treatment effect, no matter how trivial, can achieve statistical significance at any level if the sample size is large enough.

If you accept the need to formally test the null hypothesis (that is, the hypothesis that the treatment has no effect), there is a preferred alternative to significance testing. It involves estimating the magnitude of the treatment effect, called an effect-size estimate, and placing a confidence interval around this estimate (Hedges, Cooper, & Bushman, 1992; Oakes, 1986). This alternative approach can tell not only whether the null hypothesis should be rejected at a given significance level, but also whether the observed treatment effect is large enough to be considered practically important. This book adopts the approach of estimating effect-size estimates and corresponding confidence intervals.

1.4 Operationally Defining Abstract Concepts in Research

Scientific theories are composed of abstract concepts that are linked together in some logical fashion. To test hypotheses derived from theories, researchers must tie abstract concepts to concrete representations of those concepts by means of operational definitions. An operational definition specifies the operations or techniques used to measure the concept. For example, the concept "hunger" might be defined operationally as "depriving an organism of food for 24 hours." Operational definitions are the translation of an abstract concept into a concrete reality.

In a meta-analysis that investigates the same conceptual variables, researchers often use different operational definitions. For example, consider a meta-analysis on the relation between "alcohol" and "aggression" in humans (Bushman & Cooper, 1990). Even though only experimental studies of male social drinkers were included in this meta-analysis, researchers used widely different operational definitions of the concepts "alcohol" and "aggression." Although the concept "alcohol" seems simple enough to define, it was defined in a number of ways. Researchers used different types of alcohol (for example, absolute alcohol; distilled spirits such as vodka, whiskey, rum, and bourbon; beer; wine), different doses of alcohol, and different concentrations of alcohol. The concept "aggression" also was defined in a number of ways. Some researchers used physical measures of aggression (for example, giving electric shocks or noise blasts to another person, taking money away from another person), whereas other researchers used verbal measures of aggression (for example, directing verbally abusive comments to another person, evaluating another person in a negative manner).

Any single operational definition will not fully reflect the more abstract concept that it represents (Gold, 1984). In a meta-analysis, if you find the same relation between concepts, regardless of the operational definitions used in the individual studies, then your confidence in the relation increases. In fact, you might have more confidence in the findings from a meta-analysis of five studies that used different operational definitions than in the findings from a meta-analysis of 50 studies that used the same operational definitions.

1.5 Categorical (Qualitative) and Continuous (Quantitative) Variables

A variable is a qualitative or quantitative entity that can vary or take on different (at least two) values. The value of the variable is the number or label that describes the person or object of interest. In research, variables are used to represent the abstract concepts being studied. One useful distinction is between categorical and continuous variables (for example, Agresti, 1990). A categorical variable simply records which of several distinct categories or groups a person or object falls into. Some examples of categorical variables include political party affiliation, religious denomination, sex, and psychiatric diagnostic groups (for example, schizophrenia, major depression, generalized anxiety disorder). The numbers that are assigned to categorical variables are used only as labels or names; words or letters would work as well as numbers. For the variable SEX, for instance, you could assign the value 1 to males and the value 2 to females. These values do not imply that females are twice as good as males or that you could calculate the "average sex." With categorical variables, you generally calculate the number or the percent of people in each category. The values of a categorical variable are qualitatively different, whereas the values of a continuous variable are quantitatively different. Some examples of continuous variables include temperature, weight, income, and blood alcohol concentration. Mathematical operations (for example, differences, averages) make sense with continuous variables but not with categorical variables. In the SAS language, categorical variables are called classification (CLASS) variables. Variables not specified in a CLASS statement are assumed to be continuous.

1.5.1 Types of Variables in Research

1.5.1.1 Independent and Dependent Variables

Researchers generally are interested in studying the relations among two or more variables. Suppose that two variables are being studied, a stimulus (X) and a response (Y), and the researcher wants to know whether the stimulus affects the response. For example, a medical researcher might want to know whether taking aspirin (X) reduces the likelihood of a heart attack (Y), and a psychological researcher might want to know whether viewing television violence (X) increases aggression (Y). This relation between variables X and Y is depicted in Figure 1.5.

Figure 1.5 *Effects of a stimulus (X) on a response (Y)*

If the stimulus (X) can be controlled or manipulated by the researcher, it is called the independent variable (treatment or intervention). It is "independent" in the sense that its values are created by the researcher and are not affected by anything else that happens in the study. The corresponding response variable (Y) is called the dependent variable (dependent measure or outcome). It is "dependent" in the sense that its values are assumed to depend upon the values of the independent variable.

If the stimulus (X) cannot be manipulated by the researcher, it is called a predictor variable. In human participants, individual differences such as sex, age, race, religion, political affiliation, intelligence, ability, personality, risk status (for example, smoker or nonsmoker), and disease status (for example, HIV positive or negative) can be measured but cannot be (ethically) manipulated. The corresponding response variable (Y) is called the criterion variable.

In this book, X is called the independent variable or treatment, and Y is called the dependent variable or outcome, regardless of whether the researcher manipulated X. Although this usage is not technically accurate, it makes for smoother prose and it simplifies discussion considerably.

The relation between variables X and Y may be influenced by third variables. Two types of third variables, moderator variables and mediator variables, are described respectively in the next sections.

1.5.1.2 Moderator Variables

A moderator variable influences the strength and/or direction of the relation between the independent and dependent variables (Baron & Kenny, 1986). In a study by Stern, McCants, and Pettine (1982), for example, individuals were more likely to become seriously ill if they experienced uncontrollable life events (for example, death of a spouse) than if they experienced controllable life events (for example, being fired from a job). In this example, the type of life event (that is, controllable versus uncontrollable) is the moderator variable. Moderators are typically introduced when there is a weak or inconsistent relation between the independent and dependent variables (Baron & Kenny, 1986). The moderating effects of variable Z on the relation between variables X and Y is depicted in Figure 1.6. In meta-analysis, moderators are any known study characteristics that are associated with differences in effect-size estimates between studies.

Figure 1.6 *Moderating effects of the third variable (Z) on the relation between the stimulus (X) and the response (Y)*

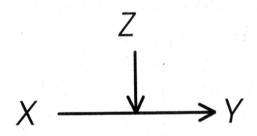

1.5.1.3 Mediator Variables

A mediator variable is the generative mechanism through which the independent variable influences the dependent variable (Baron & Kenny, 1986). Mediator variables are sometimes called intervening variables because they come between the stimulus and the response. Independent variables produce changes in mediator variables that, in turn, produce changes in dependent variables. Berkowitz (1990), for example, proposes that aversive events (for example, provocation, frustration, hot temperature) increase impulsive aggression because they produce negative affect – an unpleasant emotional response. Berkowitz views negative affect as a possible mediator between aversive events and impulsive aggression. Mediators are typically introduced when there is a strong relation between the independent and dependent variables (Baron & Kenny, 1986). The mediating effect of variable Z on the relation between variables X and Y is depicted in Figure 1.7.

Figure 1.7 Mediating effects of the third variable (Z) on the relation between the stimulus (X) and the response (Y)

$$X \longrightarrow Z \longrightarrow Y$$

1.5.2 Effect-Size Measures for Categorical Variables

Suppose that the independent and dependent variables in a study are both dichotomous (that is, both are categorical variables with two levels). For such studies, which are very common in the field of medicine, the odds ratio is the most frequently used effect-size metric. For example, Table 1.1 depicts the results from a large randomized, double-blind, placebo-controlled trial testing whether aspirin reduces mortality from cardiovascular disease (Steering Committee of the Physicians Health Study Group, 1988). The study participants, 22,071 male physicians, took either an aspirin or a placebo every other day. The data from the study at the five-year follow-up are reported here as percentages.

Table 1.1 *Results from a large randomized, double-blind, placebo-controlled trial testing whether aspirin reduces mortality from cardiovascular disease*

	Heart attack	No heart attack
Aspirin	0.94%	99.06%
Placebo	1.71%	98.29%

The odds of not having a heart attack in the aspirin group are 99.06 to 0.94 or 105.38 to 1. The odds of not having a heart attack in the placebo group are 98.29 to 1.71 or 57.48 to 1. To compare the aspirin and placebo groups, simply create a ratio of these two odds: $105.38 \div 57.48 = 1.83$. Thus, physicians in the placebo group are almost twice as likely to have a heart attack as physicians in the aspirin group. An odds ratio of 1.0 means that the aspirin doesn't differ from the placebo in reducing heart attacks. Chapter 4 discusses how to combine odds ratios.

1.5.3 Effect-Size Measures for Continuous Variables

Two measures of effect dominate the meta-analytic literature when the dependent variable is continuous: the standardized mean difference and the Pearson product-moment correlation coefficient. When the primary studies in question compare two groups, either through experimental (treatment) versus control group comparisons or through orthogonal contrasts, the effect-size estimate often is expressed as some form of standardized difference between the group means. For example, suppose that 100 participants in a study are randomly assigned to experimental or control groups. Suppose also that the mean score for the experimental group is higher ($\bar{Y}_E = 10$) than the mean score for the control group ($\bar{Y}_C = 8$), but that the variation in scores is about the same for the two groups (pooled standard deviation, $S_{POOLED} = 4$). To calculate a standardized mean difference, the control group mean is subtracted from the experimental group mean and this difference is divided by the pooled standard deviation – that is, $(10-8)/4 = 0.5$. According to Cohen (1988), a "small" standardized mean difference is 0.2, a "medium" standardized mean difference is 0.5, and a "large" standardized mean difference is 0.8. Thus, the treatment effect in our hypothetical example is medium sized.

When two continuous variables are related, the Pearson product-moment correlation coefficient (r) is most often used. Values of r can range from +1.0 (a perfect positive correlation) to −1.0 (a perfect negative correlation). A correlation coefficient of 0 indicates that the two variables are not (linearly) related. The sign on the correlation gives the direction of the relation between the two variables – a positive sign indicates that the relation is positive, whereas a negative sign indicates the relation is negative. The value of the correlation indicates the strength of the relation. Most correlations are not perfect. According to Cohen (1988), a "small" correlation is ± .1, a "medium" correlation is ± .3, and a "large" correlation is ± .5. Chapter 5 discusses how to combine standardized mean differences and correlation coefficients.

1.6 Some Issues to Consider When You Conduct a Meta-Analysis

1.6.1 Publication Bias and Study Quality

It is well documented that studies that report statistically significant results are more likely to be published than are studies reporting nonsignificant results (for example, Greenwald, 1975). In meta-analysis, the conditional publication of studies with significant results has been called the "file drawer problem" (Rosenthal, 1979). The most extreme version of this problem would result if only 1 out of 20 studies conducted was published and the remaining 19 studies were located in researchers' file drawers (or garbage cans), assuming that the .05 significance level is used. If publication bias is a problem, then the studies included in a meta-analysis may represent a biased subset of the total number of studies that are conducted on the topic. Chapter 3 describes some graphing procedures that can be used to detect publication bias.

One way to reduce publication bias is to include unpublished studies (for example, theses, dissertations) in the meta-analysis. Including unpublished studies in a meta-analysis, however, raises questions about the qualitative differences between published and unpublished studies. Because most refereed journals have reasonably strict standards for publication, published studies may be more methodologically sound than unpublished studies. Eysenck (1978) argued that when researchers fail to exclude studies of poor design, a meta-analysis becomes an exercise in "mega-silliness" that only demonstrates the axiom "garbage in — garbage out." Our personal belief is that unpublished studies should be included in a meta-analysis, but that studies should be coded on variables related to methodological quality (for example, random assignment, double blind procedures, publication status). You can then test whether the coded variables moderate the treatment effects (see Chapters 8 and 9).

1.6.2 Missing Effect-Size Estimates

Missing data is perhaps the largest problem facing the practicing meta-analyst. Missing effect-size estimates pose a particularly difficult problem because meta-analytic procedures cannot be used at all without a statistical measure for the results of a study (Pigott, 1994). Sometimes research reports do not include enough information (for example, means, standard deviations, statistical tests) to permit the calculation of an effect-size estimate. Unfortunately, the proportion of studies with missing effect-size estimates in a meta-analysis is often quite large, about 25% in psychological studies (Bushman & Wang, 1995, 1996). Vote-counting procedures can be used on studies that don't report enough information to calculate effect-size estimates but do report information about the direction and/or statistical significance of results (Bushman, 1994). Vote-counting procedures are described in Chapter 6.

Currently, the most common "solutions" to the problem of missing effect-size estimates are (a) to omit from the review those studies with missing effect-size estimates and analyze only complete cases, (b) to set the missing effect-size estimates equal to zero, (c) to set the missing effect-size estimates equal to the mean that is obtained from studies with effect-size estimates, (d) to set studies equal to the conditional mean that is obtained from studies with effect-size estimates (that is, Buck's, 1960, method), and (e) to use the available information in a research report to get a lower limit for the effect-size estimate (Rosenthal, 1994). Unfortunately, all of these procedures have serious problems that limit their usefulness (Bushman & Wang, 1996).

We proposed an alternative procedure for handling missing effect-size estimates (Bushman & Wang, 1996). Our procedure, called the combined procedure, combines sample effect-sizes and vote counts to estimate the population effect size. We believe that the combined procedure, described in Chapter 7, is the method of choice for handling missing effect-size estimates if some studies do not provide enough information to calculate effect-size estimates but do provide information about the direction and/or statistical significance of results.

1.6.3 Fixed- and Random-Effects Models

Effect-size estimates should not be combined unless they are homogeneous or similar in magnitude. You can formally test whether effect-size estimates are too heterogeneous to combine. A statistically significant heterogeneity test implies that variation in effects between-studies is significantly larger than you would expect by random chance. Between-studies variation in effects can be treated as fixed or random (Hedges & Olkin, 1985). The fixed-effects model assumes that the population effect size is a single fixed value, whereas the random-effects model assumes that the population effect size is a randomly distributed variable with its own mean and variance. When between-studies effect-size variation is treated as fixed, the only source of variation treated as random is the within-studies sampling variation. By entering known study characteristics in an analysis of variance (ANOVA) or regression model, the meta-analyst might be able to explain the "extra" variation between-studies (see Hedges, 1994). If the "extra" variation can be explained by a few simple study characteristics, then a fixed-effects model should be used. When a fixed-effects model is used, generalizations can be made to a universe of studies with similar study characteristics. The reviewers should use random-effects models if the differences between studies are too complicated to be captured by a few study characteristics. When a random-effects model is used, generalizations can be made to a universe of such diverse studies. Although generalizability is higher for random-effects models than for fixed-effects models, statistical power is higher for fixed-effects models than for random-effects models, (Rosenthal, 1995). Consequently, effect-size confidence intervals are narrower for fixed-effects models than for random-effects models. Fixed- and random-effects models are discussed in Chapters 8 and 9, respectively.

1.6.4 Correlated Effect-Size Estimates

Most meta-analytic procedures are based on the assumption that the effect-size estimates that are to be combined are independent. This independence assumption, however, is often violated. Some studies may compare multiple variants of a treatment with a common control. These studies, called multiple-treatment studies (Gleser & Olkin, 1994), will contribute more than one treatment versus control effect-size estimate. Because of the common control group, the effect-size estimates will be correlated. Other studies, called multiple-endpoint studies (Gleser & Olkin, 1994), may include only one treatment and one control but may use multiple dependent variables as endpoints for each participant. A treatment versus control effect-size estimate may be calculated for each endpoint measure. Because measures on each participant are correlated, the effect-size estimates for the measures will be correlated within studies. The best way to combine correlated effect-size estimates is to use multivariate procedures (Gleser & Olkin, 1994; Kalaian & Raudenbush, 1996). We discuss multivariate procedures in meta-analysis in Chapter 10.

1.7 Using the SAS System to Conduct a Meta-Analysis

Although meta-analytic procedures have been around for about 100 years, only since the advent of the digital computer have meta-analytic methods become accessible to practicing meta-analysts. A good meta-analytic software package should have the capability to (a) manage meta-analytical databases, (b) perform numerical calculations based on meta-analytical procedures, (c) use graphical displays to illustrate assumptions about meta-analytic procedures and to present the findings from a meta-analytical review, (d) produce the summary report of a meta-analytical review. None of the existing meta-analytic packages, however, have all of these capabilities (see Normand, 1995, for a review). Although SAS software is not specifically designed to conduct meta-analytic reviews, it has the procedures that are needed to manage databases, analyze data, and graph results. Thus, we believe that SAS is the software of choice for conducting a meta-analytic review. We hope that the SAS code in this book will make meta-analytic methods even more accessible to individuals who want to conduct a meta-analysis.

1.8 References

Bakan, D. (1966). The test of significance in psychological research. *Psychological Bulletin, 66,* 1–29.

Bangert-Drowns, R. L. (1986). Review of developments in meta-analytic method. *Psychological Bulletin, 99,* 388–399.

Baron, R. M., & Kenny, D. A. (1986). The moderator-mediator variable distinction in social psychological research: Conceptual, strategic, and statistical considerations. *Journal of Personality and Social Psychology, 51,* 1173–1182.

Berkowitz, L. (1990). On the formation and regulation of anger and aggression: A cognitive-neoassociation analysis. American Psychologist, *45,* 494–503.

Berkson, J. (1938). Some difficulties of interpretation encountered in the application of the chi-square test. *Journal of the American Statistical Association, 33,* 526–542.

Birge, R. T. (1932). The calculation of errors by the method of least squares. *Physical Review, 40,* 207–227.

Buck, S. F. (1960). A method of estimation of missing values in multivariate data suitable for use with an electronic computer. *Journal of the Royal Statistical Society, Series B, 22,* 302–303.

Bushman, B. J. (1994). Vote-counting procedures in meta-analysis. In H. Cooper & L. V. Hedges (Eds.), *The handbook of research synthesis* (pp. 193–213). New York: Russell Sage Foundation.

Bushman, B. J., & Cooper, H. M. (1990). Effects of alcohol on human aggression: An integrative research review. *Psychological Bulletin, 107,* 341–354.

Bushman, B. J. & Wang, M. C. (1995). A procedure for combining sample correlations and vote counts to obtain an estimate and a confidence interval for the population correlation coefficient. *Psychological Bulletin, 117,* 530–546.

Bushman, B. J., & Wang, M. C. (1996). A procedure for combining sample standardized mean differences and vote counts to estimate the population standardized mean difference in fixed effects models. *Psychological Methods, 1,* 66–80.

Carver, R. P. (1978). The case against statistical significance testing. *Harvard Educational Review; 48,* 378–399.

Cochran, W. G. (1937). Problems arising in the analysis of a series of similar experiments. *Journal of the Royal Statistical Society, Supplement 4*(1), 102–118.

Cochran, W. G. (1943). The comparison of different scales of measurement for experimental results. *Annals of Mathematical Statistics, 14,* 205–216.

Cohen, J. (1988). *Statistical power analysis for the behavioral sciences* (2nd ed.). Hillsdale, NJ: Lawrence Erlbaum.

Cohen, J. (1994). The earth is round ($p < .05$). *American Psychologist, 49,* 997–1003.

Cooper, H. M., & Hedges, L. V. (1994). Research synthesis as a scientific enterprise. In H. Cooper & L. V. Hedges (Eds.) *Handbook of research synthesis* (p. 4). New York: Russell Sage Foundation.

Cooper, H. M., & Rosenthal, R. (1980). Statistical versus traditional procedures for summarizing research findings. *Psychological Bulletin, 87,* 442–449.

Eysenck, H. J. (1978). An exercise in mega-silliness. *American Psychologist, 33,* 517.

Falk, R. (1986). Misconceptions of statistical significance. *Journal of Structural Learning, 9,* 83–96.

Fisher, R. A. (1932). *Statistical methods for research workers* (4th ed.). London: Oliver & Boyd. (Original work published 1925. Final edition published 1951).

Friedman, H. (1968). Magnitude of an experimental effect and a table for its rapid estimation. *Psychological Bulletin, 70,* 245–251.

Glass, G. V. (1976). Primary, secondary, and meta-analysis of research. *Educational Researcher, 5,* 3–8.

Gleser, L. J., & Olkin, I. (1994). Stochastically dependent effect sizes. In H. Cooper & L. V. Hedges (Eds.). *The handbook of research synthesis* (pp. 339–355). New York: Russell Sage Foundation.

Gold, J. A. (1984). *Principles of psychological research.* Homewood, IL: Dorsey Press.

Greenwald, A. G. (1975). Consequences of prejudice against the null hypothesis. *Psychological Bulletin, 82,* 1–20.

Harris, M. J. (1991). Significance tests are not enough: The role of effect-size estimation in theory corroboration. *Theory and Psychology, 1,* 375–382.

Hedges, L. V. (1994). Fixed effects models. In H. Cooper & L. V. Hedges (Eds.). *The handbook of research synthesis* (pp. 285–299). New York: Russell Sage Foundation.

Hedges, L. V., Cooper, H. M., & Bushman, B. J. (1992). Testing the null hypothesis in meta-analysis. A comparison of combined probability and confidence interval procedures. *Psychological Bulletin, 111,* 188–194.

Hogben, L. (1957). *Statistical theory.* London: Allen & Unwin.

Hunt, M. (1997). *How science takes stock: The story of meta-analysis.* New York: Russell Sage Foundation.

Kalaian, H. A., & Raudenbush, S. W. (1996). A multivariate mixed linear model for meta-analysis. *Psychological Methods, 1,* 227–235.

Kirk, R. E. (1996). Practical significance: A concept whose time has come. *Educational and Psychological Measurement, 56,* 746–759.

Kupfersmid, J. (1988). Improving what is published: A model in search of an editor. *American Psychologist, 43,* 635–642.

Mann, C. (1994). Can meta-analysis make policy. *Science, 266,* 960–962.

Meehl, P. E. (1978). Theoretical risks and tabular asterisks: Sir Karl, Sir Ronald, and the slow progress of soft psychology. *Journal of Consulting and Clinical Psychology, 46,* 806–834.

Morrison, D. E., & Henkel, R. E. (Eds.). (1970). *The significance test controversy.* Chicago: Aldine.

Normand, J (1995). Meta-analytical software review. *American Statistician, 49,* 352–365.

Nunnally, J. (1960). The place of statistics in psychology. Educational and *Psychological Measurement, 20,* 641–650.

Oakes, M. (1986). *Statistical inference: A commentary for the social and behavioural sciences.* New York: Wiley.

Olkin, I. (1990). History and goals. In K. W. Wachter & M. L. Straf (Eds.), *The future of meta-analysis* (pp. 3–26). New York: Russell Sage Foundation.

Pearson, K. (1904). Report on certain enteric fever inoculation statistics. *British Medical Journal, 2,* 1243–1246.

Pearson, K. (1933). On a method of determining whether a sample of size *n* supposed to have been drawn from a parent population having a known probability integral has probably been drawn at random. *Biometrika, 25,* 379–410.

Pigott, T. D. (1994). Methods for handling missing data in research synthesis. In H. Cooper & L. V. Hedges (Eds.), *Handbook of research synthesis* (pp. 163–175). New York: Russell Sage Foundation.

Rosenthal, R. (1994). Parametric measures of effect size. In H. Cooper & L. V. Hedges (Eds.) *Handbook of research synthesis* (pp. 231–244). New York: Russell Sage Foundation.

Rosenthal, R. (1979). The "file-drawer problem" and tolerance for null results. *Psychological Bulletin, 86,* 638–641.

Salsburg, D. S. (1985). The religion of statistics as practiced in medical journals. American Statistician, 39, 220–223.

Schmidt, F. L. (1996). Statistical significance testing and cumulative knowledge in psychology: Implications for training of researchers. *Psychological Methods, 1,* 115–129.

Schulman, J. L., Kupst, M. J., & Suran, B. G. (1976). The worship of "p": Significant yet meaningless research results. *Bulletin of the Menninger Clinic, 40,* 134–143.

Shaver, J. P. (1993). What statistical significance testing is, and what it is not. *Journal of Experimental Education, 61,* 293–316.

Steering Committee of the Physicians' Health Study Research Group. (1988). Preliminary report: Findings from the aspirin component of the ongoing Physicians' Health Study. *New England Journal of Medicine, 318,* 262–264.

Stern, G. S.; McCants, T. R.; Pettine, P. W. (1982). The relative contribution of controllable and uncontrollable life events to stress and illness. *Personality and Social Psychology Bulletin, 8,* 140–145.

Stouffer, S. A., Suchman, E. A., DeVinney, L. C., Star, S. A., & Williams, R. M., Jr. (1949). The American soldier: Adjustments during army life, Vol. 1. Princeton, NJ: Princeton University Press.

Tippett, L. H. C. (1931). *The methods of statistics.* London: Williams and Norgate.

Yates, F. (1951). The influence of *Statistical Methods for Research Workers* on the development of the science of statistics. *Journal of the American Statistical Association, 46,* 19–34.

Yates, F., & Cochran, W. G. (1938). The analysis of groups of experiments. *Journal of Agricultural Science, 28,* 556–580.

chapter 2

Using the SAS® System to Conduct a Meta-Analysis

2.1 Introduction

We realize that most practicing meta-analysts are not computer programmers, and some meta-analysts are not even familiar with SAS software. We have, therefore, tried to make our programs as user friendly as possible. This chapter is meant to serve as a primer for conducting a meta-analysis using SAS software. If you are familiar with SAS software, you can skip some or all of the sections in this chapter.

Section 2.2   will benefit you if you are not familiar with SAS software.

Section 2.3   describes how to create a meta-analytic SAS data set from the SAS DATA step and from a database file that supports the Open Database Connectivity (ODBC) standard (for example, Microsoft Excel).

Section 2.4   describes how you can use the SAS DATA step to rename, keep, and drop variables in an output SAS data set.

Section 2.5   describes how to create a permanent SAS data library.

Section 2.6   describes how to use SAS macros.

Section 2.7   explains how to create and manipulate a SAS graph.

Section 2.8   describes all of the SAS procedures that are used in this book.

Section 2.9   offers concluding comments.

All of the SAS data sets and SAS programs that we use in this book are available to you. See "Using This Book" for details on how to access the SAS code that we show throughout the book. "Using This Book" also explains the conventions that we use to present syntax.

2.2 Help for New Users of SAS Software

2.2.1 SAS/ASSIST Software

You can use SAS/ASSIST software to access the power of SAS software quickly and easily without having to learn SAS programming statements (SAS, 1996a, 1992). To begin, start SAS/ASSIST. The Primary Menu in the SAS/ASSIST window contains the following options (SAS, 1996a, pp. 5–8). To choose an option, simply click on it.

- **TUTORIAL.** The tutorial option provides an orientation to both SAS/ASSIST software and to the SAS System.

- **DATA MGMT.** The data management option enables you to enter new data and access existing data.

- **REPORT WRITING.** The report writing option enables you to create a variety of reports (for example, data lists, tables, charts, plots).

- **GRAPHICS.** The graphics option enables you to produce high- and low-resolution graphics (for example, bar charts, pie charts, plots).

- **DATA ANALYSIS.** The data analysis option enables you to perform several types of analyses (for example, analysis of variance, regression analysis)

- **PLANNING TOOLS.** The planning tools option enables you to analyze and compare loans, design and analyze experiments, produce quality control statistics, perform project management tasks, and forecast time series.

- **EIS.** The Executive Information System (EIS) option enables you to access SAS public applications and create and access your own applications.

- **REMOTE CONNECT.** The remote connect option enables you to establish a SAS session on a remote computer, communicate between that remote session and a local SAS session, run saved programs on the remote computer, and transfer data between the local and remote SAS sessions.

- **RESULTS.** The results option enables you to access saved programs, output, logs, and graphs.

- **SETUP.** The setup option allows you to control certain aspects of the SAS session (for example, associating reference names with data libraries or files, sorting data, reviewing function key settings).

- **INDEX.** The index option provides an alphabetical list of all of the tasks that you can perform with SAS/ASSIST software. Clicking on any task in the list takes you directly to that task.

- **EXIT.** This option is used to exit SAS/ASSIST software.

2.3 Creating a Meta-Analytic Data Set Using SAS Software

We use a real-life example to illustrate how to use SAS software. The example, from the field of medicine, is a meta-analysis on the relation between calcium intake and bone mass in young and middle-aged adults (Welten, Kemper, Post, & Van Staveren, 1995). This topic has practical significance because the bone mass that is developed earlier in life is an important determinant of osteoporotic fractures later in life. The results from 20 cross-sectional studies that reported Pearson product-moment correlation coefficients (r) are reported in Table 2.1.

Table 2.1 Welten et al. (1995) meta-analysis of the relation between calcium intake and bone mass in young and middle-aged adults

Study	n	Sex of Participants	Mean calcium intake (mg/d)	r
1	86	F	1052	.29
2	88	F	738	.00
3	60	F	871	.24
4	183	F	576	.21
5	37	F	818	.15
6	101	F	800	.10
7	300	F	909	.07
8	248	F	684	.00
9	30	F	694	−.16
10	89	F	458	.00
11	182	F	936	.08
12	161	F	456	.21
13	296	F	446	.33
14	55	F	783	.26
15	114	F	465	.19
16	173	F	436	.03
17	249	F	660	.11
18	98	F	1086	−.03
19	189	F	1310	.00
20	84	M	1313	.03

Note: M = males, F = females.

2.3.1 Creating a Meta-Analytic Data Set from the SAS DATA Step

Data must be in a form that SAS software can recognize and handle before you can use SAS to analyze them. That is, the data must be in a SAS data set. The SAS DATA step contains a set of statements that allows SAS to access and manipulate the data. When a SAS DATA step is submitted to the SAS System, it is first compiled and then executed. During the compiling phase, SAS statements are checked for errors.

Four basic types of data are used in SAS DATA steps (SAS, 1990b, pp. 20–22): (a) data from an external file, (b) data read instream, (c) data from existing SAS data sets, and (d) data generated from programming statements. Only the first three types of DATA steps are used in this book. We describe each of these three types of DATA steps in the following sections. If you are interested in learning about the fourth type of DATA step, consult the following references: SAS *Language*: *Reference* (SAS, 1990b) and SAS *Language and Procedures* (SAS, 1989).

2.3.1.1 Reading Data from an External File

The following DATA step statements are used to create a SAS data set from raw data that are stored in an external file:

```
DATA sas-data-set-name;
    INFILE "external-data-file-name";
    INPUT variable-list;
    <other-SAS-statements>
RUN;
```

The DATA statement begins the step and gives a name to the SAS data set that is being created. The INFILE statement identifies the name of the external data set. The INPUT statement gives a name to each variable in the data set and also identifies the location of each variable in the data set. If a variable in the INPUT statement is a characters string, then a $ is placed after the variable name. The RUN statement marks the end of the DATA step. Between the INPUT and RUN statements, other SAS statements can be added (for example, create new variables, modify the data).

Example 2.1 *Reading data from an external file*

The following DATA step statements provide an example of how to create a SAS data set from raw data that are stored in an external file. Suppose that the data in Table 2.1 are stored in an external file called BONE.DAT in the directory called D:\METABOOK\CH2\DATASET.

```
data ex21;
    infile "d:\metabook\ch2\dataset\bone.dat";
    input studyid n sex $ calcium r;
    z = 0.5*log((1+r)/(1-r));
    wt = n-3;
    wtz = wt*z;
run;
```

In this example, the SAS data set name is EX21. **This data set and all others used in this chapter are available via SAS Online Samples (see "Using This Book.")** The variables in the INPUT statement are STUDYID (that is, study identification number), N (that is, sample size), SEX, CALCIUM, and R (that is, Pearson product-moment correlation coefficient). The variable STUDYID corresponds with the first column of data; the variable N corresponds with the second column of data; and so on. There is a $ after the variable SEX because the values of this variable are character strings. Three SAS statements were added between the INPUT and RUN statements. The first statement computes a Fisher's (1921) z score for each correlation in the data set; the second statement computes the weight (that is, the inverse of the variance of the z score); and the third statement computes the product of the z score and its weight.

2.3.1.2 Reading Data Instream

The following DATA step statements are used to create a SAS data set from data lines read instream:

DATA *sas-data-set-name;*
    INPUT *variable-list;*
    <other-SAS-statements>
CARDS;
lines-of-raw-data
RUN;

The DATA statement begins the step and gives a name to the SAS data set that is being created. The INPUT statement gives a name to each variable in the data set and also identifies the location of each variable in the data set. The CARDS statement marks the end of the programming statements and the start of the data. Lines of raw data follow the CARDS statement. The RUN statement marks the end of the DATA step. Between the INPUT and CARDS statements, other SAS statements can be added.

Example 2.2 *Reading data instream*

The following DATA step statements provide an example of how to create a SAS data set from lines of raw data read instream:

```
data ex22;
    input studyid n sex $ calcium r;
    z = 0.5*log((1+r)/(1-r));
    wt = n-3;
    wtz = wt*z;
cards;
 1  86 F 1052   .29
 2  88 F  738   .00
 3  60 F  871   .24
 4 183 F  576   .21
 5  37 F  818   .15
 6 101 F  800   .10
 7 300 F  909   .07
 8 248 F  684   .00
 9  30 F  694  -.16
10  89 F  458   .00
11 182 F  936   .08
12 161 F  456   .21
13 296 F  446   .33
14  55 F  783   .26
15 114 F  465   .19
16 173 F  436   .03
17 249 F  660   .11
18  98 F 1086  -.03
19 189 F 1310   .00
20  84 M 1313   .03
run;
```

In this example, the data set name is EX22. The variables STUDYID, N, SEX, CALCIUM, and R in the INPUT statement correspond with the first, second, third, fourth, and fifth columns of data, respectively. The four SAS statements between the INPUT and CARDS statements are the same ones that are used in Example 2.1.

2.3.1.3 Reading Data from Existing SAS Data Sets

The following DATA step statements are used to create a SAS data set from an existing SAS data set:

DATA *new-sas-data-set-name;*
    SET *old-sas-data-set-name;*
    BY *variable-list;*
    <other-SAS-statements>
RUN;

The DATA statement begins the step and gives a name to the SAS data set that is being created. The SET statement reads observations from one or more existing SAS data sets. The optional BY statement can be used to group the data by the variables that are specified in the variable-list. To use the BY statement, the data set must be sorted in order for groups defined in the variable list. To sort the data, you should use the following statements before the DATA statement:

PROC SORT;
    BY *variable-list;*
RUN;

You must use the same variable-list in the SORT procedure and the BY statement of the DATA step. The RUN statement marks the end of the SAS DATA step. Between the BY and RUN statements, you can add other SAS statements.

Example 2.3 *Reading data from existing SAS data sets*

    Suppose that you wish to calculate a weighted average of the sample correlations (see Chapter 5, Section 5.5), r_+ (RPLUS), from a weighted average of the corresponding Fisher's z scores

$$\text{ZPLUS} = z_+ = \frac{\sum_{i=1}^{k}(n_i - 3)z_i}{\sum_{i=1}^{k}(n_i - 3)}, \tag{2.1}$$

where k is the number of studies in the meta-analysis; n_i is the sample size in the ith study; and $(n_i - 3)$ is the weight (that is, the inverse of the variance) of the Fisher's z score in the ith study. The weighting factor gives studies with smaller variances (that is, larger sample sizes) more weight. Studies with large samples should produce more accurate effect-size estimates than studies with smaller sample sizes. The weighted average correlation, r_+ (RPLUS), can then be obtained from the weighted average Fisher's z score, z_+ (ZPLUS), by means of the transformation

$$\text{RPLUS} = r_+ = \frac{\exp\{2z_+\} - 1}{\exp\{2z_+\} + 1} \tag{2.2}$$

(see Chapter 5, Section 5.5).

Suppose that the SAS data set named EX210 is an output data file from SAS PROC MEANS that contains two sums (see Section 2.7.1):

(a) $\text{WTZSUM} = \sum_{i=1}^{k} (n_i - 3)z_i$, that is, the numerator of Equation 2.1, and

(b) $\text{WTSUM} = \sum_{i=1}^{k} (n_i - 3)$, that is, the denominator of Equation 2.1. The following statements can be used to compute ZPLUS and RPLUS, respectively.

```
data ex23;
   set ex213;
   zplus = wtzsum / wtsum;
   rplus = (exp(2*zplus)-1)/(1 + exp(2*zplus));
run;
```

2.3.2 Creating a Meta-Analytic Data Set from an ODBC Data Set

Meta-analytical data sets are often created using other software packages, such as Microsoft Excel and other products that run in a Windows operating environment. In our discussion, we use Microsoft Excel as an example because of its widespread use. Microsoft Excel supports Open Database Connectivity (ODBC) standard, an interface standard that provides a common application programming interface (API) for accessing databases. SAS provides an easy way to access data that were created with other software packages that support the ODBC standard:

1. Select the "Import" option from the FILE menu in SAS.

2. Select the type of software used to create the data set (for example, Microsoft Excel) in the "Import Wizard - Select import type" menu. Press the "Next" button.

3. Type the Microsoft Excel data set name in the "Where is the file located?" field of the "Import Wizard - Select file" menu. Press the "Next" button.

4. Type the library name in the member field of "Import Wizard - Select Library and member" menu. The default SAS library name is WORK. You only need to type the member name (that is, the name of the SAS data set). Press the "Finish" button.

─────────

Example 2.4 Creating a meta-analytic data set from a Microsoft Excel file

Suppose that the data in Table 2.1 were entered in a Microsoft Excel file called EX24.XLS in the directory C:\METABOOK\CH2\DATASET. The following steps are used to create a SAS data set called EX24.SSD in the default library WORK:

1. Select the "Import" option from the FILE menu in SAS.

2. Select Microsoft Excel in the "Import Wizard - Select import type" menu. Press the "Next" button.

3. Type C:\METABOOK\CH2\DATASET\EX24.XLS in the "Where is the file located?" field of the "Import Wizard - Select file" menu. Press the "Next" button.

4. Type EX24 in the member field of the "Import wizard - Select library and member" menu. Press the "Finish" button.

The above steps produce a SAS data set called EX24 in the default library WORK.

2.4 Manipulating a SAS Data Set Using the SAS DATA Step

2.4.1 Renaming Variables

The RENAME statement in a DATA step specifies new names for variables in the output SAS data sets (SAS, 1990b, pp. 472–473). The DATA step statements are as follows:

DATA *new-sas-data-set-name*;
    SET *old-sas-data-set-name*;
        RENAME *old-variable-name=new-variable-name ...*;
RUN;

Example 2.5 *Renaming variables*

In some of the SAS macros in this book, variables have fixed names. For example, the SAS macro WAVGEFF in Appendix 5.6 uses ESTIMATE as the variable name for effect-size estimate. The following statements change the variable name Z (that is, Fisher's z score) to the variable name ESTIMATE in the meta-analysis on the relation between calcium intake and bone mass (Welten et al., 1995):

```
data ex25;
   set ex22;
   rename z=estimate;
run;
```

2.4.2 Keeping and Dropping Variables in Output SAS Data Sets

The KEEP statement is used to include the necessary variables in the output SAS data sets (SAS, 1990b, pp. 423–424), whereas the DROP statement is used to exclude the unnecessary variables from the output SAS data sets (SAS, 1990b, p. 337). You should not use the KEEP and DROP statements in the same DATA step.

If the number of variables to include is smaller than the number of variables to exclude, use the KEEP statement. The DATA step statements are as follows:

DATA *new-sas-data-set-name*;
    SET *old-sas-data-set-name*;
    KEEP *variable-name(s)*;
RUN;

If the number of variables to exclude is smaller than the number of variables to include, use the DROP statement. The DATA step statements are as follows:

DATA *new-sas-data-set-name*;
    SET *old-sas-data-set-name*;
    DROP *variable-name(s)*;
RUN;

_____

Example 2.6 *Keeping and dropping variables in output SAS data sets*

Because only one of the studies in Table 2.1 used male participants, you could exclude the variable SEX and use the following statements:

```
data ex26;
   set ex22;
   drop sex;
run;
```

The SAS data set EX26 is the same as the SAS data set EX22 except the variable SEX has been excluded from the SAS data set EX26. Note that the $ is not required after the variable SEX. After a variable has been defined as a character string variable with the $ in the input statement, the $ is not needed in subsequent SAS statements.

2.5 Creating a Permanent SAS Data Library

A SAS data library is a collection of one or more SAS files (SAS, 1989, pp. 473–484). Although a SAS data library can store several types of SAS files, we only discuss SAS data sets because they are the only types of files used in this book.

The LIBNAME statement is used to assign a libref to a SAS data library. The appropriate form of the LIBNAME statement (SAS, 1990b, pp. 431–434) is as follows:

LIBNAME *libref "your-data-library"*;

The library reference (*libref*) is the name to associate with the SAS data library. Once a libref has been established for a SAS data library, the libref can be used in a SAS program to read SAS data sets in that data library or to create new SAS data sets. "*Your-data-library*" is used to name the directory.

A file in a SAS data library can be referenced using either a one-level name (FILENAME) or a two-level name (LIBREF.FILENAME). A SAS data set referenced by a one-level name is only temporary; that is, it cannot be used in another SAS session. A SAS data set referenced by a two-level name is permanent; that is, it can be used in another SAS session.

Example 2.7 *Creating a permanent SAS data set using a two-level name*

The following SAS statements are used to create a permanent SAS data set from a temporary SAS data set:

```
libname ch2 "d:\metabook\ch2\dataset";
data ch2.ex27;
   set ex22;
run;
```

The LIBNAME statement establishes a connection between the SAS libref called CH2 and *"your-data-library"* called D:\METABOOK\CH2\ DATASET. The DATA step creates the SAS data set EX27 in the SAS data library CH2 using the two-level name CH2.EX27. The SAS data set CH2.EX27 is permanent and can, therefore, be used in different SAS sessions. For example, you can print this data set in another SAS session using the following SAS program statements (see Section 2.8.2 for a discussion of the PRINT procedure):

```
proc print data=ch2.ex27;
run;
```

2.6 Using a SAS Macro

Some meta-analytical procedures that are used in this book cannot be performed with standard SAS procedures and require special SAS macros. The SAS macro facility (SAS, 1997) is a tool for extending and customizing the SAS System and for reducing the amount of program text that you need to perform common tasks. By assigning names to groups of SAS programming statements, you can work with the names rather than with the program text.

When the SAS System compiles program text, two delimiters trigger macro processor activity: *&name* and *%name*. The *&name* refers to a macro variable, whereas the *%name* refers to a macro.

The following statement defines a macro-variable and assigns it a value:

%LET *macro-variable=value*;

The value defined in the %LET statement is substituted for the macro-variable in the printed output (SAS, 1990e, pp. 87–88). In the SAS program, the ampersand (&) is used to prefix the *macro-variable*.

This book includes several SAS macros to perform meta-analytic procedures. Some of these macros use SAS/IML (Interactive Matrix Language) software to perform matrix operations (Schoct, 1997) to fit linear models (Graybill, 1976, and Searle, 1971). You do not need to understand how to use SAS/IML, however, because the SAS/IML procedures are used within SAS macros. You only need to know how to use SAS macros.

This section describes how to use SAS macros but not how to write them. If you want to write your own SAS macros, refer to *SAS Guide to Macro Processing* (SAS, 1990a) and *SAS Macro Language* (SAS, 1997). The following steps are required to use the SAS macros in this book:

1. Select the "Open" option from the "File" menu in SAS.
2. Select the name of the macro from the directory it is in. Press the "Open" button to include the macro in the PROGRAM EDITOR window.
3. Select the "Submit" option from the "Locals" menu to compile the macro.

   Type the following SAS statements in PROGRAM EDITOR window:

   ```
   <other-SAS-statements>
   %macro-name(<variable-list>);
   <other-SAS-statements>
   RUN;
   ```

1. Select the "Submit" option from the "Locals" menu to produce the output from the SAS macro in the OUTPUT window.
2. The following example illustrates how to use one of the SAS macros in this book.

Example 2.8 Using the SAS macro WAVGMETA in Appendix 5.6

Suppose that a meta-analyst wants to use the SAS macro WAVGMETA (*indata, outdata, kind, level*) in Appendix 5.6 to compute a weighted average of sample correlations in Table 2.1 (for more details see Chapter 5, Section 5.5). Suppose that the SAS macro WAVGMETA is in the directory D:\META BOOK\CH5\SASCODE. The following steps explain how to use the macro:

1. Select the "Open" option from the "File" menu in SAS.

2. Select SAS MACRO WAVGMETA in the directory D:\METABOOK\CH5\SASCODE. Press the "Open" button to include the macro in the PROGRAM EDITOR window.

3. Select the "Submit" option from the "Locals" menu to compile the macro.

4. Type the following SAS statements in PROGRAM EDITOR window:

```
options nocenter pagesize=54 linesize=80 pageno=1;
libname ch2 "d:\metabook\ch2\dataset";
data ch2.ex28;
    set ex22;
    type = 4;
    ne = 0.5 * n;
    nc = n - ne;
    rename z = estimate;
%wavgmeta(ch2.ex28,ch2.ex28out,4,.95);
proc format;
    value efftype 0 = "Glass' Delta"
                  1 = "Cohen's d  "
                  2 = "Hedges' g  "
                  3 = "Hedges' gu "
                  4 = "Correlation ";
proc print data=ch2.ex28out noobs;
    title;
    var type estimate lower upper;
    format type efftype. estimate lower upper 5.3;
run;
```

In line 9, the specific parameters are given for the SAS macro WAVGMETA(*indata,outdata,kind,level*). CH2.EX2X1 is the name of the input data file, *indata*; CH2X1OUT is the name of the output data file, *outdata*; 4 indicates that the kind of effect-size estimate to be combined is the correlation coefficient; and .95 is the level of confidence for the confidence interval (that is, a 95% confidence interval). A "%" is used in line 9 because WAVGMETA is a macro name rather than a variable name.

- Select the "Submit" option from the "Locals" menu to produce the following output in the OUTPUT window:

TYPE	ESTIMATE	LOWER	UPPER
Correlation	0.117	0.080	0.153

As you can see in the OUTPUT window, the estimated correlation is $r_+ = .117$ with a 95% confidence interval ranging from .080 to .153. The confidence interval does not include the value zero, suggesting that there is a positive relation between calcium intake and bone mass.

2.7 Creating and Manipulating a SAS Graph

In this book, SAS/GRAPH software is used to produce funnel plots that use the SAS macro FUNNEL and normal quantile plots that use the SAS macro CIQQPLOT (see Chapter 3, Sections 3.3 and 3.4, respectively). In this section, we describe how to use these macros. For a complete discussion of SAS/GRAPH software, consult *SAS/GRAPH Software Usage* (1991) and *SAS/GRAPH Software Reference* (1990c).

2.7.1 SAS/GRAPH Displays on the Computer Monitor

It is often useful to see a graph on a computer monitor before it is printed (to save both time and trees!). The following steps create a SAS/GRAPH funnel plot or a normal quantile plot on a computer monitor (see Section 2.6 for a discussion on how to use SAS macros).

1. To make a funnel plot, include the SAS macro FUNNEL in Appendix 3.4 in the PROGRAM EDITOR window. To make a normal quantile plot, include the SAS macro CIQQPLOT in Appendix 3.6 in the PROGRAM EDITOR window.

2. Select the "Submit" option from the "Locals" menu to compile the macro.

3. Type the following in the PROGRAM EDITOR window:

```
LIBNAME libref <your-data-library>;
GOPTIONS <options>;
    <other-SAS-statements>
    %macro-name(<variable-list>);
QUIT;
```

4. Select the "Submit" option from the "Locals" menu to produce the graph file on the SAS/GRAPH window.

You can examine the plot on the computer monitor, print the plot, or copy the plot to a document, but you cannot manipulate features of the plot (for example, font size of text). The next section describes how to use a computer graphics metafile (CGM) to produce a graph file that can be manipulated using other software packages (for example, Microsoft PowerPoint).

Example 2.9 SAS/GRAPH displays on the computer monitor

The following steps create a normal quantile plot, displayed on the computer monitor, for the data in Table 2.1:

1. Include the SAS macro CIQQPLOT in Appendix 3.6 in the PROGRAM EDITOR window.

2. Select the "Submit" option from the "Locals" menu to compile the macro.

3. Type the following SAS code in the PROGRAM EDITOR window:

```
libname ch2 "d:\metabook\ch2\dataset";
goptions reset=goptions gsfname=fig201
   gsfmode=replace gunit=pct;
data temp;
   set ch2.ex27;
   sz = z / sqrt(1/(n-3));
%ciqqplot(temp,ch2graph,fig201,sz);
quit;
```

In line 6, the specific parameters are given for the SAS macro CIQQPLOT(*indat,graphlab,cgmfile,yvarname*). TEMP is the name of the input data file, *indat*; CH2GRAPH is the output graphical category name, *graphlab*; FIG201 is the output CGM graphical file name, *cgmfile;* and SZ is the name of the variable on the *Y*-axis, *yvarname*. SZ, calculated in line 5, is the standardized effect-size estimate that is obtained by dividing the Fisher's *z* score by its estimated standard deviation.

4. Select the "Submit" option from the "Locals" menu to produce the normal quantile plot on the SAS/GRAPH window.

2.7.2 Creating a SAS/GRAPH CGM File

A computer graphics metafile (CGM) file can be manipulated by most existing
software packages, such as Microsoft PowerPoint. For example, you can change
the size, type, and style of text font in a CGM file. You can use the following steps
to create a SAS/GRAPH CGM file:

1. To make a funnel plot CGM file, include the SAS macro FUNNEL in
   Appendix 3.4 in the PROGRAM EDITOR window. To make a normal
   quantile plot CGM file, include the SAS macro CIQQPLOT in Appendix 3.6
   in the PROGRAM EDITOR window.

2. Select the "Submit" option from the "Locals" menu to compile the macro.

3. Type the following in the PROGRAM EDITOR window:

   LIBNAME *libref <your-data-library>*;
   FILENAME *fileref <filename>*;
   GOPTIONS *<options>*;
      <other-SAS-statements>
      %macro-name(<variable-list>);
   QUIT;

4. Select the "Submit" option from the "Locals" menu to produce the CGM file.

Example 2.10 Creating a SAS/GRAPH CGM file

The following steps create a normal quantile plot CGM file for the data in
Table 2.1:

1. Include the SAS macro CIQQPLOT in Appendix 3.6 in the PROGRAM
   EDITOR window.

2. Select the "Submit" option from the "Locals" menu to compile the macro.

3. Type the following SAS code in the PROGRAM EDITOR window:

```
libname ch2 "d:\metabook\ch2\dataset";
filename fig202 "c:\tmp\fig202.cgm";
goptions reset=goptions device=cgmmwwc gsfname=fig202
   gsfmode=replace ftext=hwcgm005 gunit=pct;
data temp;
   set ch2.ex27;
   sz = z / sqrt(1/(n-3));
%ciqqplot(temp,ch2graph,fig202,sz);
quit;
```

Line 1 is needed because the data set is stored in the SAS data library called D:\METABOOK\CH2\DATASET (see Section 2.5 for a discussion of SAS data libraries). Line 2 gives the name of the CGM file (that is, FIG202.CGM) and specifies its directory location (that is, C:\TMP). The only thing that you have to change in the GOPTIONS statement is the name of the plot in GSFNAME=<*plot-name*>.

4. Select the "Submit" option from the "Locals" menu to produce the CGM file FIG202.CGM in the directory C:\TMP.

You can then copy and paste this CGM file into other applications that support the CGM standard, such as Microsoft PowerPoint and Microsoft Word. All of the figures in this book were manipulated using Microsoft PowerPoint and Microsoft Word.

2.8 SAS Procedures Used in This Book

In this book, we use the following SAS procedures: SORT, PRINT, MEANS, UNIVARIATE, GLM, FORMAT, TIMEPLOT, SHEWHART, MIXED, GPLOT, and IML. In this section we describe only the first eight procedures. We do not describe the MIXED procedure because it is only used to estimate the random effects variance in Chapter 9. To learn more about the MIXED procedure, consult *SAS/STAT Software: Changes and Enhancements* (SAS, 1996b). We do not describe the GPLOT and IML procedures in this section because you do not need to modify any PROC GPLOT or PROC IML statements (see Section 2.6).

For more information on the GLM procedure, consult *SAS/STAT User's Guide* (SAS, 1990f). For more information on the GPLOT procedure, consult *SAS/GRAPH Software: Usage* (1991a) and *SAS/GRAPH Software: Reference* (1990d). For more information on the IML procedure, see *SAS/IML Software: Usage and Reference* (SAS, 1990e). For more information on the other SAS procedures, consult *SAS Procedures Guide* (SAS, 1990c).

2.8.1 PROC SORT

The SORT procedure sorts observations in a SAS data set by one or more variables. The sorted observations either replace the original data set or are stored in a new SAS data set. In this section, we describe the PROC SORT statements that are used in this book (SAS, 1990c, pp. 505–513):

PROC SORT <*option-list*>
    BY <DESCENDING>*variable-1*<...<DESCENDING>*variable-n*>;
RUN;

The PROC SORT statement starts the procedure. Of the available options in the *option-list*, we only use DATA=*sas-data-set*, which specifies the name of the data set to be sorted. The BY statement specifies the variables to be sorted on. Character variables are sorted in ascending alphabetical order (that is, from letter A to Z), whereas continuous variables are sorted in ascending numerical order (that is, from the smallest value to the largest value). To sort the observations in descending order, use the keyword DESCENDING before the name of each variable in the BY statement whose values you want sorted in descending order.

Example 2.11 *Illustration of the SORT procedure*

It is often useful to sort effect-size estimates in descending order. You can use the following SAS code to sort the correlations in Table 2.1 in descending order. The SAS data set CH2.EX27, used to illustrate all the SAS procedures in this section, was created in Example 2.7:

```
proc sort data=ch2.ex27;
   by descending r;
run;
```

2.8.2 PROC PRINT

The PRINT procedure prints all or some of the observations in a SAS data set. In this section, we describe the PROC PRINT statements that are used in this book (SAS, 1990c, pp. 461–482):

PROC PRINT <*option-list*>;
    VAR *variable-list*;
    BY *variable-list*;
RUN;

The PROC PRINT statement starts the procedure. Of the available options in *option-list*, we only use DATA=*sas-data-set* and NOOBS. The DATA=*sas-data-set* option specifies the name of the data set to be printed. The NOOBS option suppresses the observation number in the printed output. If the VAR statement is deleted, all of the variables in the data set are printed. The BY statement, described in Section 2.8.1, is optional for PROC PRINT. If the BY statement is used, the print procedure expects the input data set to be sorted in the order of the BY variables. You can use the SORT procedure to sort the input data set (see Section 2.8.1). The RUN statement executes the previously entered SAS statements.

Example 2.12 *Illustration of the PRINT procedure*

You can use the following SAS code to print the observations in Table 2.1:

```
data ex212;
    infile "d:\metabook\ch2\dataset\bone.dat";
    input studyid n sex $ calcium r;
    z = 1/2*log((1+r)/(1-r));
    wt = n-3;
    wtz = wt*z;
run;
proc print data=ex212 noobs;
    title;
    var studyid n sex calcium r;
run;
```

If the VAR statement in this example were deleted, the PRINT procedure also would print observations for the variables Z, WT, and WTZ.

2.8.3 PROC MEANS

The MEANS procedure computes descriptive statistics for continuous variables such as sample sizes and effect-size estimates. In this section, we describe the PROC MEANS statements that are used in this book (SAS, 1990c, pp. 366–388):

PROC MEANS *<option-list>* *<statistic-keyword-list>*;
    VAR *variable-list*;
    CLASS *variable-list*;
    WEIGHT *variable*;
    BY *variable-list*;
    OUTPUT *<out=sas-data-set><output-statistic-list>*;
RUN;

The PROC MEANS statement starts the procedure. Of the available options in *option-list*, we use only DATA=*sas-data-set* and NOPRINT. The DATA=*sas-data-set* option specifies the name of the data set to be analyzed by the MEANS procedure. The NOPRINT option suppresses all printed output. Of the available statistics in *statistic-keyword-list*, we use only N (the number of observations with nonmissing values for the variable), MEAN (the mean or average scores), MIN (the minimum score), MAX (the maximum score), VAR (the variance of the scores), STD (the standard deviation of the scores), and SUM (the sum of scores). If no statistics are specially requested, PROC MEANS prints the variable name, N, MEAN, STD, MIN, and MAX. Statistics are calculated for each quantitative variable that is listed in the VAR statement. You can use the CLASS statement to obtain descriptive statistics for different groups of variables. The CLASS statement is similar to the BY statement except that the printed output differs slightly and the BY statement requires the data to be sorted first. The WEIGHT statement specifies a numeric variable in the input data set whose values are used to weight each observation. The BY statement, described in Section 2.8.1, is optional for PROC MEANS.

The OUTPUT statement outputs statistics to a new SAS data set. OUT=<*sas-data-set*> specifies the name of the output data set.

<*output-statistic-list*> specifies the type of statistics that are contained in the new data set and also specifies the names of the variables that contain these statistics. The RUN statement executes the previously entered SAS statements.

Example 2.13 *Illustration of MEANS procedure*

You can use the MEANS procedure to calculate the sums in the numerator (that is, WTZSUM) and denominator (that is, WTSUM) of Equation 2.1 for the data in Table 2.1. The SAS code is as follows. The NOPRINT option is used because the output statistics WTSUM and WTZSUM do not need to be printed.

```
proc means data=ex22 noprint;
    var wt wtz;
    output out=ex213 sum = wtsum wtzsum;
run;
```

2.8.4 PROC UNIVARIATE

The UNIVARIATE procedure, like the MEANS procedure, calculates simple descriptive statistics for continuous variables. But the UNIVARIATE procedure also calculates quantiles and extreme scores. If you use the PLOT option, the UNIVARIATE procedure produces stem-and-leaf plots, box plots, and normal probability plots. In this section, we describe the PROC UNIVARIATE statements that are used in this book (SAS, 1990c, pp. 617–834):

PROC UNIVARIATE <*option-list*>;
    VAR *variable-list*;
    BY *variable-list*;
    OUTPUT <OUT=*sas-data-set*><*output-statistic-list*>;
RUN;

    The PROC UNIVARIATE statement starts the procedure. Of the available options in *option-list*, we use the DATA= *sas-data-set* and PLOT options. The DATA= *sas-data-set* option specifies the name of the data set to be analyzed by the UNIVARIATE procedure. The PLOT option produces a stem-and-leaf plot, a box plot, and a normal probability plot for the continuous variables that are listed in *variable list*. The optional BY and OUTPUT statements are described in Sections 2.8.1 and 2.3, respectively. The RUN statement executes the previously entered SAS statements.

Example 2.14 *Illustration of UNIVARIATE procedure*

    You can use the following SAS code to obtain descriptive statistics and plots for the Fisher's z scores in the meta-analysis on the relation between calcium intake and bone mass (Welten et al., 1995):

```
proc univariate data=ch2.ex27 plot;
   var z;
run;
```

2.8.5 PROC GLM

The <u>G</u>eneral <u>L</u>inear <u>M</u>odel (GLM) procedure uses the method of generalized least squares to fit linear models such as analysis of variance (ANOVA) and regression analysis. In this section, we describe the PROC GLM statements that are used in this book (SAS, 1990f, pp. 893–996):

PROC GLM <*option-list*>;
    CLASS *variables*;
    WEIGHT *variable*;
    MODEL *dependents = independents* ;
    BY *variables*;
    OUTPUT <*out=sas-data-set*><*output-statistic-list*>;
RUN;

The PROC GLM statement starts the procedure. Of the available options in *option-list*, we use only DATA= *sas-data-set* and NOPRINT. The DATA=*sas-data-set* option specifies the name of the data set to be analyzed by the GLM procedure. The NOPRINT option suppresses all printed output.

The CLASS statement identifies the categorical independent variables that are to be included in the analysis. All independent variables not identified in the CLASS statement are assumed to be continuous (see Chapter 1, Section 1.5 for a discussion of the distinction between categorical and continuous variables). If there are no categorical independent variables, the CLASS statement is not used. The WEIGHT statement specified a variable for weighting the observations. In meta-analysis, effect-size estimates are frequently weighted by the inverse of their variance.

The continuous dependent variables to be analyzed are listed on the left side of the MODEL statement, whereas the continuous and categorical independent variables lists are listed on the right side of the MODEL statement (see Chapter 1, Section 1.5.1.1 for a discussion of the distinction between independent and dependent variables). The optional BY and OUTPUT statements are described in Sections 2.8.1 and 2.3, respectively. The RUN statement executes the previously entered SAS statements.

Example 2.15 Illustration of GLM procedure

The following SAS code tests whether the Fisher's z scores are homogeneous in the meta-analysis on the relation between calcium intake and bone mass (Welten et al., 1995). The homogeneity test statistic is the corrected total sum of squares from the GLM procedure, which has a chi-square distribution with $20-1 = 19$ degrees of freedom (see Chapter 8, Section 8.5.2).

```
proc glm data=ch2.ex27;
   weight wt;
   model z = calcium;
run;
```

2.8.6 PROC FORMAT

The FORMAT procedure defines informats and formats for character and numeric variables (SAS, 1990c, pp. 275–312). Because informats are not used in this book, we only describe formats. Formats tell the SAS System the data's type (that is, character or numeric) and its form (for example, leading zeros, decimal and comma punctuation). The FORMAT procedure creates two kinds of formats: value and picture formats. Because picture formats are not used in this book, we only describe value formats. Value formats convert output values to a different form, either numeric to character (for example, the value 1 to "Males" and the value 2 to "Females") or character string to a different character string (for example, the character "M" to "Males" and the character "F" to "Females").

The FORMAT procedure contains the following statements:

PROC FORMAT *<option-list>*;
   VALUE *name <(format-option-list)>*
   range-1='formatted-value-1'
   <...range-n='formatted-value-n'>;
RUN;

Example 2.16 *Illustration of FORMAT procedure*

The following SAS code reformats the values of the SEX variable for the meta-analysis on calcium intake and bone mass (see Table 2.1):

```
proc format;
    value $sexfmt "M" = "Males" "F" = "Females";
data ex16;
    infile "d:\metabook\ch2\dataset\bone.dat";
    input studyid n sex $ calcium r;
    format sex $sexfmt.;
run;
```

Had the variable SEX been coded 1 for males and 2 for females, the following SAS code would have been used instead:

```
proc format;
    value sexfmt 1 = 'Males' 2 = 'Females';
data ex16;
    infile "d:\metabook\ch2\dataset\bone.dat";
    input studyid n sex $ calcium r;
    format sex sexfmt.;
end;
```

2.8.7 PROC TIMEPLOT

The TIMEPLOT procedure plots one or more variables over time intervals. In this book, we use the TIMEPLOT procedure to create dot plots (see Chapter 3, Section 3.2). A dot plot displays an effect-size estimate and 95% confidence interval separately for each study. Generally, it is helpful to rank order studies by the effect-size estimate rather than by the first author's last name. If the effect-size estimates are related to a study characteristic, then the effects should be rank ordered within levels of the study characteristic. In this section, we describe the PROC TIMEPLOT statements that are used in this book (SAS, 1990c, pp. 579–593):

PROC TIMEPLOT *<option-list>*;
    PLOT *request-list </ option-list>*;
    CLASS *variable-list*;
    BY *variable-list*;
RUN;

The PROC TIMEPLOT statement starts the procedure. Of the available options in *option-list*, we only use the DATA= *sas-data-set* option. The PLOT statement specifies the plots to produce. The *request-list* specifies the variables to plot and, optionally, the symbol to mark the points on the plot. Suppose that you want the effect-size estimate (EFFECT) to be marked with an asterisk "*," and you want the lower (LOWER) and upper (UPPER) 95% confidence interval bounds to be marked with square brackets "[" and "]", respectively. The plot statement would be: `plot lower="[" effect="*" upper="]"`.

Of the available options in *</ option-list>*, we use OVERLAY, HILOC, REF=*<value>*, and REFCHAR=*< value >*. We use 0 as the value in the latter two options. The OVERLAY option overlays all variables that are specified in the PLOT statement onto one plot. The HILOC option connects the lower bound to the upper bound of the 95% confidence interval with a line of hyphens (that is, ----). The REF=0 option draws a line on the plot perpendicular to the value 0 on the horizontal axis. The REFCHAR=0 option specifies the character 0 as the symbol for drawing the reference line.

The optional CLASS and BY statements were described in Sections 2.8.5 and 2.8.1, respectively. The ID statement prints the values of the ID variables in the listing but does not plot them. We use the variable STUDY in the ID statement. The RUN statement executes the previously entered SAS statements.

***Example 2.17** Illustration of TIMEPLOT procedure*

The following SAS code produces a dot plot for the studies in Table 2.1. LOWER and UPPER are the variables for the lower and upper 95% confidence interval bounds, respectively. The formulas to obtain the effect-size estimates and 95% confidence intervals are given in Chapter 5. The results are printed in Output 2.1.

```
options nodata nocenter pagesize=54 linesize=80 pageno=1;
data ch2.ex217;
    set ch2.ex27;
    std = sqrt(1 / (n-3));
    lower = z + probit(0.025)*std;
    upper = z + probit(0.975)*std;
    lower = (exp(2*lower)-1)/(exp(2*lower)+1);
    upper = (exp(2*upper)-1)/(exp(2*upper)+1);
proc sort data=ch2.ex217;
    by r;
proc timeplot data=ch2.ex217;
    title;
    plot lower="[" r="*" upper="]"/
    overlay hiloc ref=0 refchar="0";
    id studid;
run;
```

Of the 20 studies shown in Output 2.1, five studies had significant positive correlations and 15 studies had confidence intervals that included the value zero.

Output 2.1 *Dot plot for the individual studies in the meta-analysis on calcium intake and bone mass*

2.8.8 PROC SHEWHART

The SAS SHEWHART procedure produces high resolution graphics output (see Chapter 25 in *SAS/QC Software: Usage and Reference*, 1995). In this book, we use PROC SHEWART to make high-resolution side-by-side box plots (see Chapter 3, Section 3.6). Side-by-side box plots are especially useful when the magnitude of effect-size estimates is related to a categorical moderator variable. Separate box plots are constructed for each category of the moderator variable, and the box plots are placed side-by-side so the reader can compare the effect-size estimate distributions for different categories. In this section, we describe the PROC SHEWART statements that are used in this book.

PROC SHEWART *<option-list>*;
    BOXCHART *(processes)*subgroup-variable </options>*
BY *variable-list*;
RUN;

The PROC SHEWART statement starts the procedure. Although PROC SHEWART can make several different kinds of charts, in this book we use only the BOXCHART option. The *processes* is the effect-size estimate variable and the *subgroup-variable* is the categorical moderator variable. Of the available BOXCHART options, we use only BOXSTYLE=SCHEMATIC.

For schematic boxplots, we use the following arguments and values: (a) IDSYMBOL=DOT, (b) VREF=0, (c) HAXIS=AXIS1, (d) NOLEGEND, (e) HOFFSET=5, (f) NOLIMIT, and (g) STDDEVS. The IDSYMBOL=DOT argument identifies outlying observations with the dot symbol. The VREF=0 argument draws a horizontal line at the value 0. The HAXIS=AXIS1 argument defines the horizontal axis and its values. The NOLEGEND option suppresses the printing of legend. The HOFFSET=5 argument specifies the offset length in percent screen units at both ends of the horizontal axis by 5%. The NOLIMIT and STDDEVS options suppress the display of control limits. Control limits are not used in side-by-side box plots. The BY statement is described in Section 2.8.1.

Example 2.18 *Illustration of SHEWART procedure*

The following SAS code produces male and female side-by-side boxplots for the studies in Table 2.1. Because only one study actually used female participants, we modified the data set. Suppose that studies 1–10 used female participants and studies 11–20 used male participants. The results are printed in Figure 2.1. For details on how to interpret the box plots, see Chapter 3, Section 3.6.

```
/* side-by-side box plots;              */
options nodate pagesize=54 linesize=80 nocenter pageno=1;
data fig201;
   set ex22;
   if (_n_ > 10) then sex="M";
   label z="Effect Size Estimate";
   label sex="Gender";
proc sort data=fig201;
   by sex;
filename fig201 "d:\tmp\fig201.cgm";
goptions reset=goptions device=cgmmwwc gsfname=fig201
   gsfmode=replace ftext=hwcgm005 gunit=pct;
proc shewhart data=fig201 gout=fig201 graphics;
   boxchart z*sex / boxstyle=schematic
                    idsymbol=dot
                    vref=0
                    haxis=axis1
                    nolegend
                    hoffset=5
                    nolimits
                    boxwidth=30
                    stddevs;
axis1 value = ("Male" "Female");
run;
quit;
```

Figure 2.1 *Side-by-side box plots of sex differences in the relation between calcium intake and bone mass*

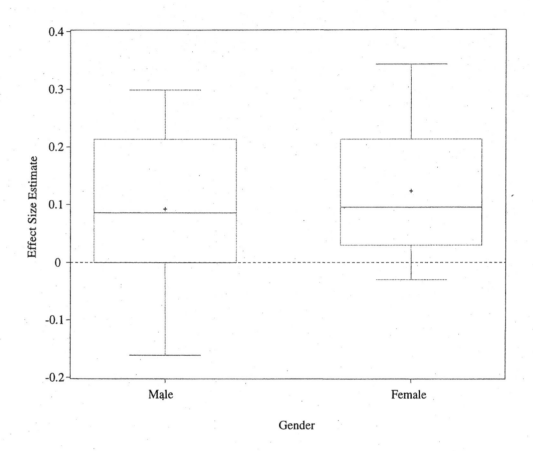

2.9 Conclusions

This chapter is meant to serve as a primer to conducting a meta-analysis using SAS software. To use the SAS programs in this book, the data from the meta-analysis must be in a SAS data set. If the data set is already saved in some other format, it can usually be converted to a SAS data set. Most of the meta-analytic procedures in this book are in macros. We tried to provide enough background information in this chapter so that readers will feel comfortable modifying the SAS code in the macros to fit their own needs.

2.10 References

Fisher, R. A. (1921). On the "probable error" of a coefficient of correlation deduced from a small sample. *Metron, 1*, 1–32.

Gilmore, J. (1996). *Painless Windows 3.1: A beginner's handbook for SAS users.* Cary, NC: SAS Institute Inc.

Graybill, F. A. (1976). *Theory and application of the linear model.* Pacific Grove, CA: Wadsworth & Brooks.

SAS Institute Inc. (1989). *SAS language and procedures: usage,* version 6, 1st ed. Cary, NC: Author.

SAS Institute Inc. (1990a). *SAS guide to macro processing,* version 6, 2nd ed. Cary, NC: Author.

SAS Institute Inc. (1990b). *SAS language: reference,* version 6, 1st ed., Cary, NC: Author.

SAS Institute Inc. (1990c). *SAS procedures guide,* version 6, 3rd ed., Cary, NC: Author.

SAS Institute Inc. (1990d). *SAS/GRAPH software: Reference,* version 6, 1st ed., volumes 1 & 2, Cary, NC: Author.

SAS Institute Inc. (1990e). *SAS/IML software: Usage and reference,* version 6, 1st ed., Cary, NC: Author.

SAS Institute Inc. (1990f). *SAS/STAT user's guide,* version 6, 4th ed., volumes 1 & 2, Cary, NC: Author.

SAS Institute Inc. (1991). *SAS/GRAPH software: Usage,* version 6, 1st ed., Cary, NC: Author.

SAS Institute Inc. (1992). *Doing more with SAS/ASSIST software,* version 6, 1st ed., Cary, NC: Author.

SAS Institute Inc. (1995). *SAS/QC software: Usage and reference*, version 6, 1st ed., volumes 1 & 2, Cary, NC: SAS Institute Inc.

SAS Institute Inc. (1996a). *Getting started with the SAS system using SAS/ASSIST software,* version 6, 2nd ed., Cary, NC: Author.

SAS Institute Inc. (1996b). *SAS/STAT software: Changes and enhancements* through release 6.11. Cary: SAS Institute Inc.

SAS Institute Inc. (1997). *SAS macro language:* reference, 1st ed., Cary, NC: Author.

Schott, J. R. (1997). *Matrix analysis for statistics.* New York: John Wiley & Sons.

Searle, S. R. (1971). *Linear models.* New York: John Wiley & Sons.

Welten, D. C., Kemper, H. C. G., Post, G. B., & Van Staveren, W. A. (1995). A meta-analysis of calcium intake in young and middle-aged females and males. *Journal of Nutrition, 125,* 2802–2813.

chapter 3

Graphical Presentation of Meta-Analytic Results

3.1 Introduction

Most people have heard the phrase "A picture is worth a thousand words." This phrase seems especially applicable to the presentation of results from a meta-analysis. The reader should not have to wade through a quagmire of data to discover important findings (Light, Singer, & Willett, 1994). If the reader cannot readily interpret the findings from a meta-analysis, then the meta-analysis is essentially worthless. An appropriate graph can go a long way toward conveying the important findings of a meta-analysis. In a meta-analysis, you can use graphical displays (figures, plots) to enhance numerical analyses in at least two ways. First, you can use graphical displays to discover patterns and relations among variables in a meta-analysis. Second, you can use graphical displays as supplementary tools to check statistical assumptions on which numerical analyses are based. Graphical displays are not, however, designed to replace more formal inferential statistical procedures.

It seems paradoxical that many authors use state-of-the-art procedures in conducting a meta-analysis, yet they still present their findings using display methods that were popular before the age of computers, graphics, and exploratory data analysis (Light, Singer, & Willett, 1994). In this chapter, we discuss how you can use SAS software to display the results of a meta-analysis in an effective and comprehensible manner. SAS code is given for five types of plots: dot plots, funnel plots, normal quantile plots, stem-and-leaf plots, and box plots.

3.2 Dot Plots

Often, a meta-analyst needs to display effect-size estimates separately for each study. This can be effectively accomplished using a dot plot (Cleveland, 1985). The studies can be identified by the first author's last name and publication year. Generally, it is helpful to rank order studies by the effect-size estimate rather than by the first author's last name. If the effect-size estimates are related to a study characteristic, then the effects should be rank ordered within levels of the study characteristic. It is also quite useful to include a confidence interval with each effect-size estimate. With a confidence interval, the reader can determine how reliable the estimate is and whether the confidence interval includes the value zero. The formulas to obtain the effect-size estimates and 95% confidence intervals are given in Chapter 5.

Output 3.1 gives an example of a dot plot for a meta-analysis on the relation between student ratings of teaching and student performance on the final exam (Cohen, 1983). A positive correlation indicates that students who did well on the final exam gave the teacher higher ratings than students who did poorly. As you can see in Output 3.1, nine studies (Bolton et al., 1979; Centra, 1977b; Elliot & Richard, 1977; Bryson, 1974; Remmer et al., 1949; Crooks & Smock, 1974; Sullivan & Skanes, 1974b; Marsh et al., 1956; Hoffman, 1978) had confidence intervals that did not include the value zero. The correlations were greater than zero in these nine studies. The results in Output 3.1 could also be displayed in a table, but a dot plot is much more effective, especially as the number of studies included in the meta-analysis grows.

Output 3.1: *Dot plot for Cohen's (1983) meta-analysis on the relation between student ratings of teaching and student performance on the final exam*

STUDY	LOWER	EFFECT	UPPER	min -0.600	max 0.917
Greenwood et al. 1976	-0.42	-0.11	0.23	[------------*----0---------]	
Doley & Crichton 1978	-0.60	-0.04	0.55	[[----------------------*-0---------------------]	
Wherry 1952	-0.30	0.16	0.56	[-------------0------*----------------]	
Frey et al. 1975	-0.44	0.18	0.68	[-----------------0---------*---------------]	
Centra 1977a	-0.37	0.23	0.69	[--------------0----------*---------------]	
McKeachie et al. 1971	-0.09	0.26	0.55	[---0---------*-----------]	
Hoffman 1978	0.05	0.27	0.47	0 [----------*--------]	
Elliot 1950	0.00	0.33	0.59	[--------------*----------]	
Sullivan & Skanes 1974c	-0.19	0.34	0.72	[--------0-------------*--------------]	
Marsh et al. 1956	0.24	0.40	0.54	0 [-------*-----]	
Sullivan & Skanes 1974b	0.10	0.40	0.63	0 [------------*----------]	
Sullivan & Skanes 1974d	-0.14	0.42	0.78	[------0-------------*-------------]	
Crooks & Smock 1974	0.14	0.49	0.73	0 [-------------*---------]	
Doyle & Whitely 1974	-0.12	0.49	0.83	[----0------------------*---------------]	
Remmer et al. 1949	0.20	0.49	0.70	0 [------------*--------]	
Sullivan & Skanes 1974a	-0.03	0.51	0.82	[0-------------------*------------]	
Bryson 1974	0.16	0.56	0.80	0 [-----------------*----------]	
Elliot & Richard 1977	0.17	0.58	0.82	0 [----------------*----------]	
Centra 1977b	0.30	0.64	0.84	0 [---------------*--------]	
Bolton et al. 1979	0.09	0.68	0.92	0 [------------------------*---------]	

The dot plot in Output 3.1 was created using the TIMEPLOT procedure in SAS (see Chapter 2, Section 2.8.7). The SAS data set CH3.COHEN83.SSD, which was created using the SAS program file COHEN83.SAS, contains four variables: STUDY, EFFECT, LOWER, and UPPER. The STUDY variable contains the information about the first author's last name and the publication year. The EFFECT variable contains the effect-size estimate for the study. The LOWER, and UPPER variables contain the lower and upper 95% confidence interval bounds, respectively. The SAS code for Output 3.1 is given in Appendix 3.1.

3.3 Funnel Plots

A *funnel plot* is a two-dimensional scatter plot with sample size on one axis and effect-size estimate on the other axis.[1] Funnel plots are often used to investigate whether all studies come from a single population and to search for publication bias (Light & Pillemer, 1984). Both uses are discussed in this section.

3.3.1 Using a Funnel Plot to Investigate Whether All Studies Come from a Single Population

In a funnel plot, the sample sizes are plotted against their respective effect-size estimates. The funnel plot capitalizes on the well-known statistical principle that sampling error decreases as sample size increases. If the studies all come from a single population, then the plot should look like a funnel with the width of the funnel decreasing as sample size increases. This is an important use of funnel plots because if all the studies come from a single population, then it makes sense to average the sample effect sizes to estimate the true population effect size.

---

[1]In this chapter, we use Hedges' *g* effect-size estimate (see Chapter 5, Section 5.2.1). Of course, a funnel plot can also be constructed using effect-size estimates other than standardized mean differences (for example, correlation coefficients).

You can use the GPLOT procedure in SAS/GRAPH to create a funnel plot (see Chapter 2, Section 2.7). This book contains a SAS macro called FUNNEL that is based on the GPLOT procedure (see Appendix 3.3). If you are an experienced SAS/GRAPH user, invoke the macro using the SAS code directly. If you are a beginning SAS/GRAPH user, invoke the macro by entering parameters on the prompted window displays. Consult Chapter 2, Section 2.6, before using the SAS macro FUNNEL if you are not an experienced SAS/GRAPH user. Additional resources include Chapter 31 of *SAS/GRAPH Software* (SAS, 1990a) and TS-252E, *Exporting SAS/GRAPH Output to Microsoft PowerPoint for Windows* (SAS, 1994).

The funnel plot in Figure 3.1 depicts a simulation data set of 120 studies that compared two groups (for example, experimental versus control). The data were simulated to have a common variance 1 for both groups. The mean difference between two groups was treated as a fixed number 0. That is, the fixed-effects model was assumed (fixed-effects models are discussed in Chapter 8).

Three steps are required to create the funnel plot in Figure 3.1:

1.   Create a SAS data set. For Figure 3.1, we used the simulation data set DATA 301.SSD, which was created using the SAS program file DATA301.SAS. You can access the SAS program file DATA301.SAS via SAS Online Samples (see "Using This Book"). The SAS data set should contain the variables EFFECT and SAMPLE, where EFFECT is the effect-size estimate for the study, and SAMPLE is the sample size for the study.[2]

2.   Use the SAS macro FINDXY(*indat,xx,yy*) in Appendix 3.2 to compute the maximum and the minimum values for the variables on the X and Y axes (that is, EFFECT and SAMPLE), where *indat* is the input data file name, *xx* is the variable name on X-axis, and *yy* is the variable name on Y-axis.

---

[2] If the effect-size estimate is the standardized mean difference, we use the average sample size based on the square mean root method as the value of SAMPLE (see Hedges & Olkin, 1985). If the effect-size estimate is Pearson product-moment correlation coefficient, we use the total sample size as the value of SAMPLE.

3. Use the SAS macro FUNNEL in Appendix 3.4 to create Figure 3.1. The graphical output FIG301.CGM is in the computer graphics metafile (CGM) format. CGM files can be manipulated with most existing text editors, such as Microsoft PowerPoint. Use the following command to access the SAS macro FUNNEL:

%FUNNEL (*indat, xvar, xlab, xmin, xmax, xby, yvar, ylab, ymin, ymax, yby, gdir, ghout*);

where

indat is the input data file name.

xvar is the name of the variable on the X-axis.

xlab is the label on the X-axis.

xmin is the minimum value for the variable on the X-axis.

xmax is the maximum value for the variable on the X-axis.

xby is the number of tick marks on the X-axis.

yvar is the name of the variable on the Y-axis.

ylab is the label on the Y-axis.

ymin is the minimum value for the variable on the Y-axis.

ymax is the maximum value for the variable on the Y-axis.

yby is number of tick marks on the Y-axis.

gdir is the directory name for the graphical output file.

ghout is the name of the graphical output file.

The SAS macro FUNNELIN (see Appendix 3.3) can be called within the SAS macro FUNNEL to help the user input parameters that are needed for the FUNNEL macro. The user simply needs to follow the directions on the window displays to input the macro parameters that are needed for the funnel plot. To use the SAS Macro FUNNELIN, type %FUNNELIN.

Display 3.1 asks the user to input the name of the SAS data set.

Display 3.1 Input name of SAS data set

Display 3.2 asks the user to provide the name of the variable on the X-axis (for example, EFFECT or SAMPLE).

Display 3.2 *Input name of variable on X-axis*

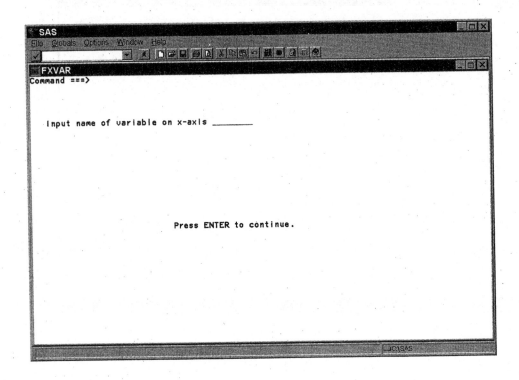

Similar displays prompt the user to provide additional parameters that are needed to produce a funnel plot. The last display prompts the user to enter the name of the graphical output file.

_____

Display 3.3 *Input name of graphical output file*

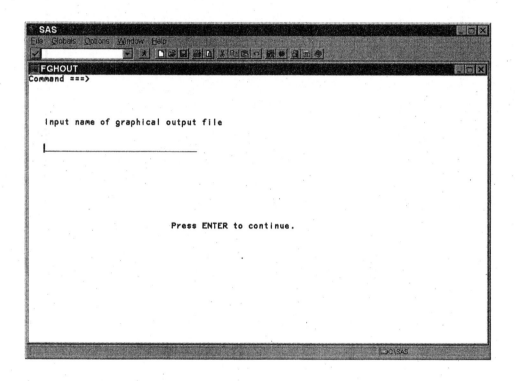

Figure 3.1 Funnel plot for simulated data set with mean difference 0 and common variance 1 (120 studies)

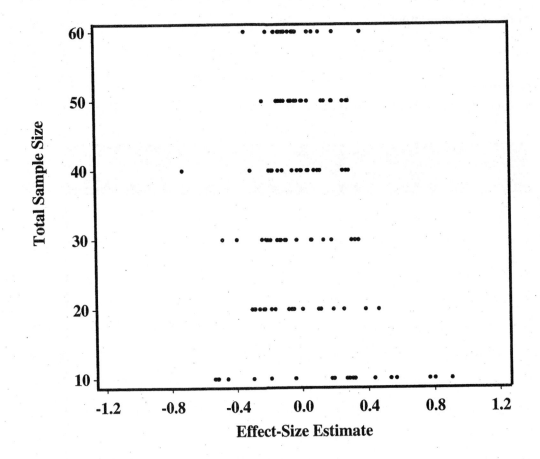

Note that as sample size increases, the width of the funnel plot decreases. This funnel plot suggests that the 120 studies do come from a single population. However, more formal inferential procedures should also be used to determine whether the effect-size estimates are homogeneous (see Chapters 8 and 9).

If studies come from two or more populations, the funnel plot should not converge to a single value. For example, Figure 3.2 shows a funnel plot that uses a simulation data set of 120 studies. Half of the studies were simulated to have a mean difference of 0, and the other half were simulated to have a mean difference of 0.8. The common variance was set at 1 for all studies. Note that as the sample size increases the width of the funnel plots does not converge to a single value. This funnel plot suggests that the 120 studies do not come from the same population because the effect-size estimates do not converge on a single value as sample size increases. For purposes of illustration, different symbols were used for effect-size estimates from different populations in Figure 3.2. In practice, however, you may not know the number of populations that were sampled or which effect-size estimates come from which populations.

You should also formally test whether the effect-size estimates are homogeneous (see Chapters 8 and 9). It is worth noting that although the homogeneity test is useful in determining whether the studies come from a single population, it is not very useful in determining how many populations the studies come from. The funnel plot can help you decide how many populations the studies come from if the population effect sizes are quite different. However, it is difficult to determine how many populations the studies come from in Figure 3.2 because the population effect sizes are not different enough. The SAS code for creating the SAS data set DATA302.SSD and Figure 3.2 is given in Appendix 3.5.

Figure 3.2 *Funnel plot for simulated data set from two populations (120 studies). The mean differences for populations 1 and 2 were set at 0 and 0.8, respectively. The variance was set at 1 for both populations.*

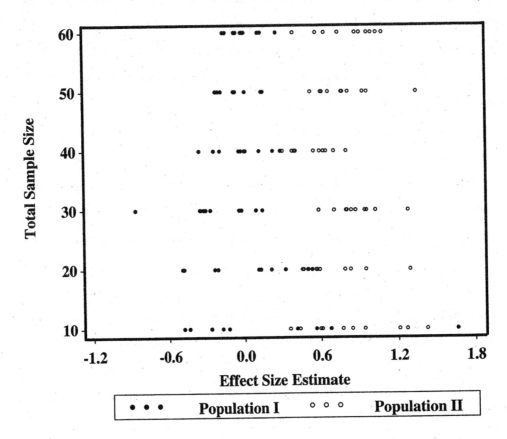

3.3.2 Using a Funnel Plot to Search for Publication Bias

A funnel plot can identify publication bias. It is well documented that studies that report statistically significant results are more likely to be published than are studies reporting nonsignificant results (for example, Greenwald, 1975). In meta-analysis, the conditional publication of studies with significant results has been called the "file drawer problem" (Rosenthal, 1979). The most extreme version of this problem would result if only 1 in 20 studies conducted was published and the remaining 19 were located in researchers' file drawers (or garbage cans).

If publication bias is a problem, then the studies that are included in a meta-analysis may represent a biased subset of the total number of studies that were conducted on the topic.

The shape of a funnel plot can suggest whether publication bias exists. If the true effect size is zero, a few studies will still have significant results — those with very large effect-size estimates (positive or negative) and those with very large sample sizes. Studies with small effect-size estimates and small sample sizes will have nonsignificant results. If publication bias exists, and the true effect is zero, then the middle of the funnel plot will be "hollow." For example, the data plotted in Figure 3.3, were obtained from a simulation study in which the mean difference was set at 0 and studies with nonsignificant results (at the .05 level) were deleted. Note that there are no effect-size estimates in the center of the funnel plot. In Figure 3.3, it is assumed that positive and negative significant results are published. When the primary study tests are directional, only half of Figure 3.3 will be seen.

Figure 3.3 *Funnel plot for simulated data set with mean difference 0 and common variance 1. Studies with nonsignificant results at the .05 level were deleted from the data set.*

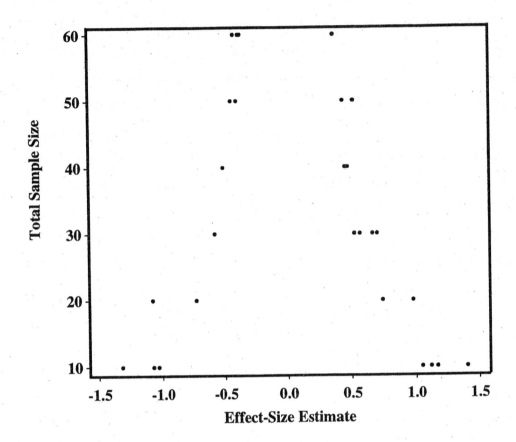

If the true effect size differs from zero, then publication bias shows up in a different form on the funnel plot. Studies with small effect-size estimates and small sample sizes will still have nonsignificant results and will not appear in the plot. If publication bias exists, and the true effect is not zero, then there should be a "bite" out of the funnel where sample sizes and effect-size estimates are small. For example, the data plotted in Figure 3.4 were obtained from a simulation study in which the population effect size was set at 0.5, and studies with nonsignificant positive results at the .05 level were deleted. Note that there is a "bite" out of the lower left hand corner of the plot where both sample sizes and effect-size estimates are small.

Figure 3.4 Funnel plot for simulated data set with mean difference 0.5 and common variance 1. Studies with nonsignificant results at the .05 level were deleted from the data set.

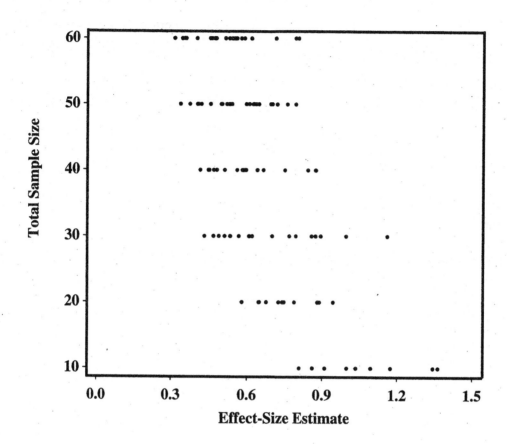

If publication bias exists, unpublished studies (for example, master's theses, doctoral dissertations) should report smaller effect-size estimates than published studies. A real-world illustration of the use of funnel plots to detect publication bias is shown in Figure 3.5. The data were taken from a meta-analytic review of 54 studies of the effect of psycho-educational interventions on post surgical hospital stay (Devine & Cook, 1983). Overall, psycho-educational interventions reduced hospital stay by about three days. However, you can see in the left corner of Figure 3.5, unpublished studies had smaller effects than did published studies, suggesting the presence of publication bias.

Figure 3.5 Funnel plot for Devine and Cook's (1983) meta-analysis of the effects of psycho-educational interventions on post surgical hospital stay

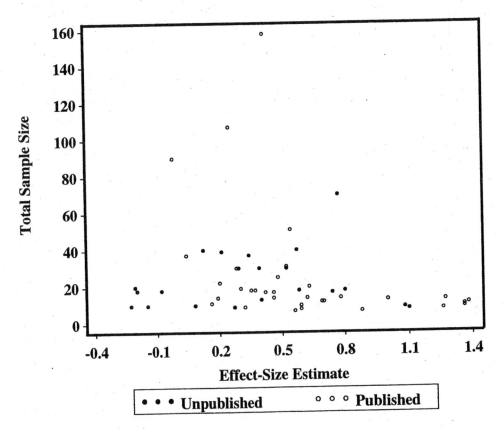

3.3.3 Problems with Funnel Plots

Unfortunately, there are three problems that limit the usefulness of funnel plots. First, it is very difficult to determine (using only your eyeballs) whether the data are shaped like a funnel, especially when the number of studies included in a meta-analysis is small. For example, Figure 3.6 shows a funnel plot that uses a simulation data set of 15 studies that compared the means from two groups. It is very difficult to determine whether the data in Figure 3.6 are shaped like a funnel.

Figure 3.6 *Funnel plot for simulated data set with mean difference 0 and common variance 1 (15 studies)*

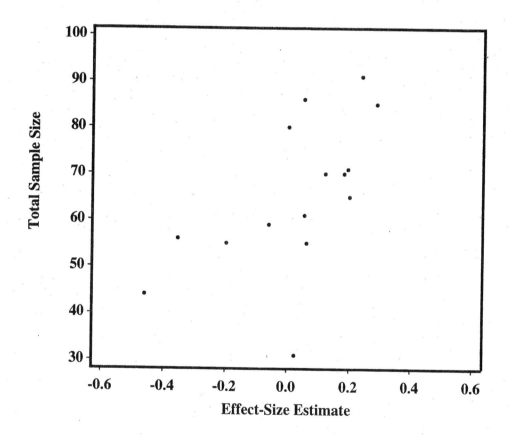

In contrast, it is much easier to determine whether the data fall on a straight line in a normal quantile plot (see the description in Section 3.4), especially if the plot includes 95% confidence interval bands. For example, Figure 3.7 uses the same simulation data set as Figure 3.6. It is easy to see that the data in Figure 3.7 fall on a straight line and within the 95% confidence bands.

Figure 3.7 Normal quantile plot for simulated data set with mean difference 0 and common variance 1 (15 studies)

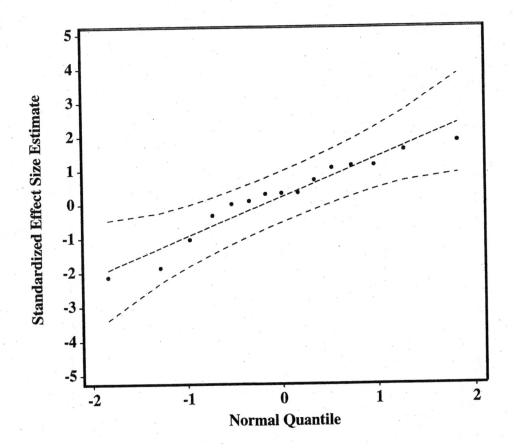

A second problem with funnel plots is that they ignore the important fact that the effect-size estimate in each study in a meta-analysis has an approximately normal distribution if the study has a large enough sample size. Most meta-analytical procedures are based on the assumption of normality (or asymptotic normality). It is, therefore, important to check this assumption before applying meta-analytic procedures that assume normality.

A third problem with funnel plots is that the data can look like a funnel even if the studies come from more than one population when the populations have the same mean but different variances. For example, Figure 3.8 shows a funnel plot that uses a simulation data set of 120 studies. Half of the studies were simulated to have a mean difference of 0 and a common variance of 1; the other half were simulated to have a mean difference of 0 and a common variance of 4. Note that as the sample size increases the width of the funnel plot decreases and converges to the value zero. The funnel plot suggests that 120 studies come from a single population with a mean difference of zero. This conclusion is wrong, however, because the studies come from two different populations.

Figure 3.8 *Funnel plot for simulated data set from two populations. Both populations have mean difference 0, but the variance of one population is 1 and the variance of the other population is 4.*

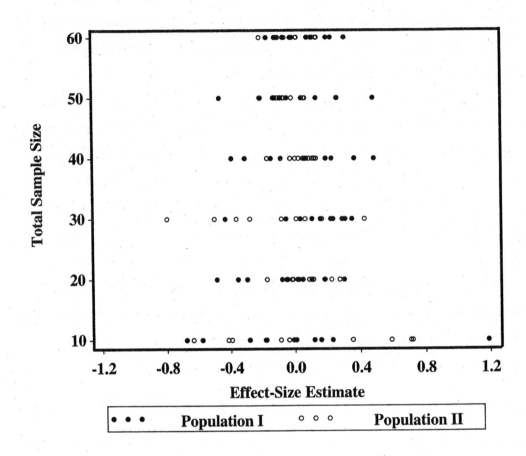

3.4 Normal Quantile Plots

Wang and Bushman (1998) proposed the normal quantile plot as a preferred alternative to the funnel plot. In a *quantile-quantile plot*, two distributions are compared by plotting their quantiles (or percentiles) against each other. The quantiles for one distribution are plotted on the *X* axis and the quantiles for the other distribution are plotted on the *Y* axis. If the two distributions are similar, then the quantiles will also be similar and the points will fall close to the line $X = Y$. Any deviation from the line $X = Y$ reveals how the distributions differ.

In a *normal quantile plot*, the quantiles of an observed distribution are plotted against the quantiles of the standard normal distribution (that is, the normal distribution with mean 0 and standard deviation 1). If the observed data have a standard normal distribution, the points on the plot will fall close to the line $X = Y$. The slope of the plotted line is equal to the standard deviation of the observed distribution, and the Y coordinate of the center point is equal to the mean of the observed distribution.

In a meta-analysis, the observed distribution consists of the effect-size estimates from a set of studies. Each standardized effect-size estimate (that is, effect-size estimate divided by its corresponding estimated standard deviation) should have a standard deviation of 1. According to the central limit theorem, the distribution of effect-size estimates will be approximately normal if the sample size in each study is large enough (say at least 30 in each study). Thus, the plotted line in the normal quantile plot should be fairly straight and should have a slope close to 1 if the studies in the meta-analysis come from a single population and if the sample size for each study is large enough.

The SAS CAPABILITY procedure can produce high-quality normal quantile plots without confidence interval bands (see Chapter 10 in *SAS/QC Software: Usage & Reference*, 1995). For this book, we wrote a macro to produce high-quality normal quantile plots with 95% confidence interval bands using the SAS GPLOT procedure. The SAS macro CIQQPLOT(*indat, graphlab, cgmfile, yvarname*), found in Appendix 3.6, was used to create all of the normal quantile plots that are presented in this chapter. *Indat* is the input data file name; *graphlab* is the output graphical category name; *cgmfile* is the output CGM graphical file name; and *yvarname* is the name of the variable on the Y-axis of the normal quantile plot. Several formulae can be used to approximate the standard error estimate that is used in constructing the 95% confidence interval bands (for example, Chambers, Cleveland, Kleiner, & Tukey, 1983; Kendall & Stuart, 1977). We used the formula in Chambers et al. (1983, p. 229).

In a meta-analysis, you can use the normal quantile plot to (a) check the normality assumption, (b) investigate whether all studies come from a single population, and (c) search for publication bias. We discuss each of these uses in turn.

3.4.1 Using a Normal Quantile Plot to Check the Normality Assumption

Figure 3.9 shows a normal quantile plot of the same simulation data set as was used for Figure 3.1. The mean difference was set at 0 and the common variance was set at 1. Note that the data depart from linearity at both ends of the curve, especially the lower end. The curve is U-shaped, suggesting that the distribution is skewed to the right. Thus, the normality assumption appears to be violated in this data set.

Figure 3.9 *Normal quantile plot for simulated data set with mean difference 0 and common variance 1 (120 studies)*

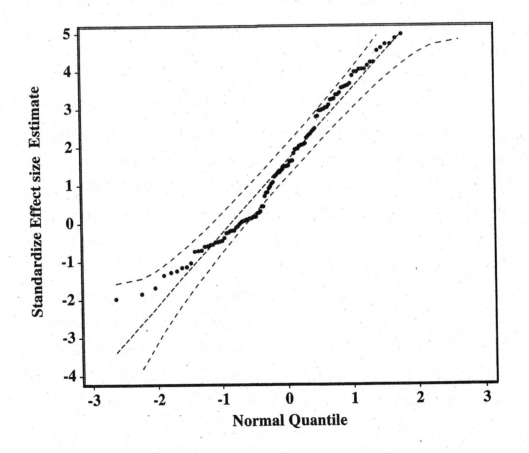

Figure 3.10 shows a normal quantile plot that uses a subset of the simulation data set that was used for Figure 3.1. Studies with sample sizes ≤ 30 were excluded from the plot. Note that the data fall on a fairly straight line and only a few points fall on the left side of the plot outside of the 95% confidence interval bands. Figure 3.10 suggests that the normality assumption is valid for studies with large sample sizes. This pattern of results is what would be expected based on the central limit theorem. In cases where the data are not normal, a data transformation might help (see Tukey, 1977, for a discussion of data transformations).

Figure 3.10 *Normal quantile plot for simulated data set with population mean difference 0 common variance 1 (60 studies). Studies with sample sizes ≤30 were deleted from the data set.*

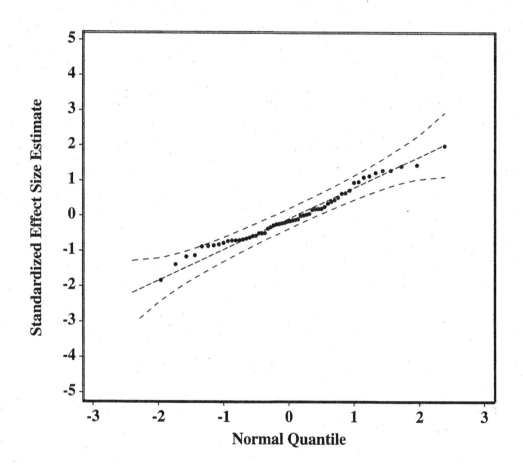

3.4.2 Using a Normal Quantile Plot to Investigate Whether All Studies Come from a Single Population

Figure 3.11 shows a normal quantile plot that uses the same simulation data set as was used for Figure 3.2. The mean difference was set at 0 for half of the studies, and at 0.8 for the other half. The common variance was set at 1 for both groups of studies. The curve in Figure 3.11 is S-shaped and has one "bump" below and one "bump" above the center dashed line, reflecting the fact that the 120 studies come from two different populations. Figure 3.11 is also easier to interpret than Figure 3.2. From Figure 3.11, it is relatively easy to determine that the studies come from two populations because there are two bends in the curve. From Figure 3.2, it is difficult to determine how many populations the studies come from.

Figure 3.11 *Normal quantile plot for simulated data set from two populations (120 studies). The effect sizes for populations 1 and 2 were set at 0 and 0.8, respectively. The common variance was set at 1 for both populations.*

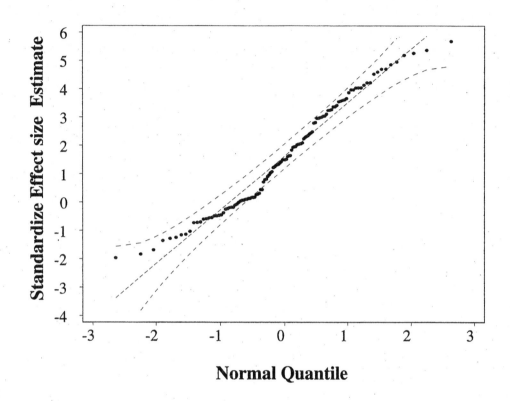

3.4.3 Using a Normal Quantile Plot to Search for Publication Bias

Figure 3.12 shows a normal quantile plot that uses the same simulation data set as was used for Figure 3.3. The studies were simulated to have a mean difference of 0, and studies with nonsignificant results at the .05 level were deleted. The fitted curve on Figure 3.12 has a suspicious gap around zero where there are no data. This gap is due to the fact that studies with nonsignificant effects were deleted. Because the gap is located around $Y = 0$, the true population effect size is very close to zero. Figure 3.12 suggests the presence of the type of publication bias that exists when the population effect size equals zero.

Figure 3.12 Normal quantile plot for simulated data set with mean difference 0 and common variance 1. Studies with nonsignificant results at the .05 level were deleted from the data set.

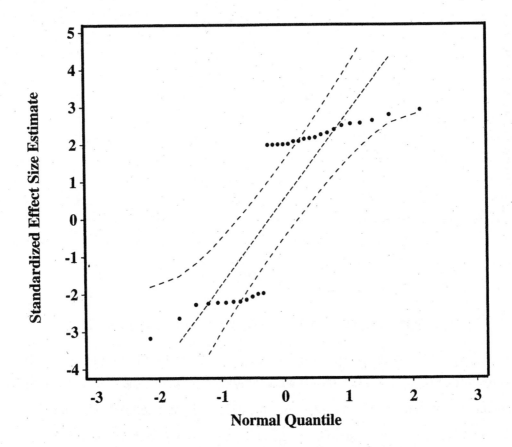

Figure 3.13 shows a normal quantile plot that uses the same simulation data set as was used as for Figure 3.4. The studies were simulated to have a mean difference of 0.5, and studies with nonsignificant results at the .05 level were deleted. Figure 3.13 shows another type of departure from linearity. The curve is

U-shaped, reflecting the fact that the data have an asymmetric distribution that is skewed to the right. The long right tail is due to the fact that studies with nonsignificant results were deleted. Figure 3.13 suggests the presence of the type of publication bias that exists when the population effect size differs from zero.

Figure 3.13 *Normal quantile plot for simulated data set with mean difference 0.5 and common variance 1. Studies with nonsignificant results at the .05 level were deleted from the data set.*

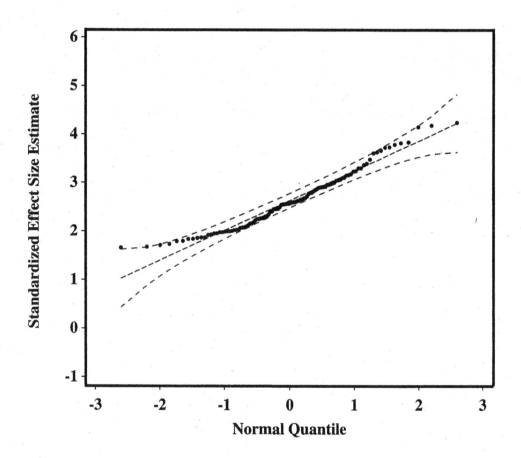

3.4.4 Problems with Normal Quantile Plots

Like all other statistical plots, normal quantile plots have weaknesses. For example, if the studies come from more than one population with means that are very close in size, it may be difficult to determine how many populations the studies come from using a normal quantile plot. In general, graphical procedures are not intended to replace the formal statistical tests that are designed for the same purpose. However, the use of graphical displays can enhance the use of statistical tests and can be used to check the assumptions on which they are based.

3.5 Stem-and-Leaf Plots

Tukey (1977) developed two useful techniques for summarizing data graphically: stem-and-leaf plots and box plots. Stem-and-leaf plots are described in this section and box plots are described in Section 3.6. A *stem-and-leaf plot* shows the shape of the distribution while including the actual numerical values in the plot. After the effect-size estimates are rank ordered in a list, the stems are separated from the leaves. The stems consist of the initial digits and are placed to the left of a vertical line; the leaves consist of the remaining digits and are placed to the right of a vertical line. Once a distribution has been displayed by a stem-and-leaf plot, you can locate the center of the distribution, examine the overall shape of the distribution, and look for marked deviations from the overall shape (for example, gaps in the distribution, outliers). No information is lost with a stem-and-leaf plot because the raw effect-size estimates can be reproduced from the plot. The stem-and-leaf plot has become a useful component of standard data summary procedures such as the UNIVARIATE procedure in SAS (SAS, 1990b; see Chapter 2, Section 2.8.4).

Output 3.2 shows a stem-and-leaf plot for 20 studies that investigated whether student ratings of teaching were related to student performance on the final exam (Cohen, 1983). The stem-and-leaf plot was produced by the UNIVARIATE with the PLOT option. The UNIVARIATE procedure also gives a box plot and other descriptive statistics such as moments and quantiles. The SAS code that produces Output 3.2 is given here. The data set contains the variable EFFECT for study effect-size estimates.

```
options nodate pagesize=54 linesize=80 nocenter pageno=1;
libname ch3 "d:\metabook\ch3\dataset";
proc univariate data=ch3.cohen83 plot;
   var effect;
run;
```

Output 3.2 *The UNIVARIATE procedure display for Cohen's (1983) meta-analysis*

```
Univariate Procedure

Variable=EFFECT

                          Moments

N                    20    Sum Wgts          20
Mean              0.364    Sum             7.28
Std Dev        0.210073    Variance    0.044131
Skewness       -0.70651    Kurtosis    0.213382
USS              3.4884    CSS          0.83848
CV             57.71227    Std Mean    0.046974
T:Mean=0       7.749021    Pr>|T|        0.0001
Num ^= 0             20    Num > 0           18
M(Sign)               8    Pr>=|M|       0.0004
Sgn Rank            102    Pr>=|S|       0.0001

                    Quantiles(Def=5)

100% Max         0.68         99%          0.68
 75% Q3           0.5         95%          0.66
 50% Med          0.4         90%          0.61
 25% Q1         0.245         10%          0.06
  0% Min        -0.11          5%        -0.075
                               1%         -0.11

Range            0.79
Q3-Q1           0.255
Mode             0.49

Stem Leaf                        #  Boxplot
    6 48                         2     |
    5 168                        3  +------+
    4 002999                     6  *------*
    3 34                         2  |  +   |
    2 367                        3  +------+
    1 68                         2     |
    0                                  |
   -0 4                          1     |
   -1 1                          1     |
      ----+----+----+----+
       Multiply Stem.Leaf by 10**-1
```

In Output 3.2, the stem represents the leading digits of the correlation coefficient. For example, the stem labeled "6" is the leading digits for correlations that range from 0.60 to 0.69. In the PROC UNIVARIATE display, each correlation is rounded to the next digit after the stem, and that digit is printed to the right of the stem. The top line appears as follows:

6   48

The line represents the sample correlations .64 and .68. The scale of the values in the stem-and-leaf plot is noted as "Multiply Stem.Leaf by 10**-1" in Output 3.2. For example, $6.4 \times 10^{-1} = .64$.

The stem-and-leaf plot depicted in Output 3.2 is unimodal. A small gap in the plot where the sample correlation is zero might suggest publication bias. As you can see in the stem-and-leaf plot, the distribution is slightly skewed to the left. This observation is confirmed by the negative skewness value.

3.6 Box Plots

A box plot gives a quick impression of certain prominent features of the distribution. It incorporates measures of location to study the variability of observations and the concentration of observations in the quantiles of the distribution. Lines are drawn at the 25th, 50th, and 75th percentiles. The 50th percentile is the median; the 25th and 75th percentiles are the first and third quartiles, respectively. The first and third quartile lines are connected to form a box that encloses the median. The distance between the third and first quartiles is called the "interquartile range." Central vertical lines, called "whiskers," extend from the box as far as the data extend, to a distance of at most 1.5 interquartile ranges. The length of whiskers relative to the box shows how long the tails of the distribution are. Extreme scores, or outliers, also are depicted in a box plot. Mild outliers are between 1.5 and 3 interquartile ranges from the box. Extreme outliers are more than 3 interquartile ranges from the box.

The box plot allows a partial assessment of symmetry of the distribution. The distribution is symmetric if (a) the median cuts the box in half, (b) the whiskers are about the same length, and (c) any outliers are symmetrically placed beyond the lower and upper whiskers (Chambers et al., 1983).

Often, the magnitude of effect-size estimates is related to systematic differences between studies, such as study characteristics. If this is the case, parallel or side-by-side box plots are quite useful. Effect-size estimates are divided into subgroups and a separate box plot is constructed for each subgroup. The box plots are placed side-by-side so the reader can see how the study characteristic is related to the magnitude of the effect-size estimate.

In a meta-analysis on alcohol-related aggression, for example, Bushman and Cooper (1990) analyzed four types of mean differences. A brief description of the four types of comparisons may be helpful. When researchers study the effects of alcohol (and other drugs) on behavior, they sometimes use a balanced placebo design (Ross, Krugman, Lyerly, & Clyde, 1962). In the balanced placebo design, half of the participants are told that they will receive alcohol and half are told that they will not receive alcohol. Within each of these groups, half of the participants are given alcohol and half are not (see Figure 3.14). The pure pharmacological effects of alcohol on aggression (uncontaminated by expectancy effects) can be determined by comparing the antiplacebo group with the control group. The pure expectancy effects of alcohol on aggression (uncontaminated by pharmacological effects) can be determined by comparing the placebo group with the control group.

Figure 3.14 Balanced placebo design

	Receive alcohol	**Don't receive alcohol**
Expect alcohol	Alcohol	Placebo
Don't expect alcohol	Antiplacebo	Control

In the laboratory, aggression is measured by having participants give noxious physical (for example, electric shocks, noise blasts) or verbal (for example, hostile comments, negative evaluations) stimuli to another person or by having participants take away positive stimuli (for example, money) from another person.

The SAS SHEWHART procedure makes high-quality side-by-side box plots (see Chapter 2, Section 2.8.8). The following SAS code was used to construct side-by-side box plots for the four types of comparisons analyzed by Bushman and Cooper (1990). The results are displayed in Figure 3.15.

```
/*  side-by-side box plots;                        */
options nodate pagesize=54 linesize=80 nocenter pageno=1;
libname ch3 "d:\research\metabook\ch3\dataset";
data fig315;
   set ch3.bac90;
   label effect="Effect Size Estimate";
   label type="Type of Comparison";
proc sort data=fig315;
   by type;
filename fig315 "d:\tmp\fig315.cgm";
goptions reset=goptions device=cgmmwwc gsfname=fig315
   gsfmode=replace ftext=hwcgm005 gunit=pct;
proc shewhart data=fig315 gout=fig315 graphics;
   boxchart effect*type / boxstyle=schematic
                          idsymbol=dot
                          vref=0
                          haxis=axis1
                          nolegend
                          hoffset=5
                          nolimits
                          stddevs.;
axis1 value =
   ("Alcohol vs. Control" "Alcohol vs. Placebo"
   "Antiplacebo vs. Control" "Placebo vs. Control");
run;
quit;
```

Figure 3.15 Side-by-side box plots produced by the SAS SHEWHART procedure for the Bushman & Cooper (1990) meta-analysis

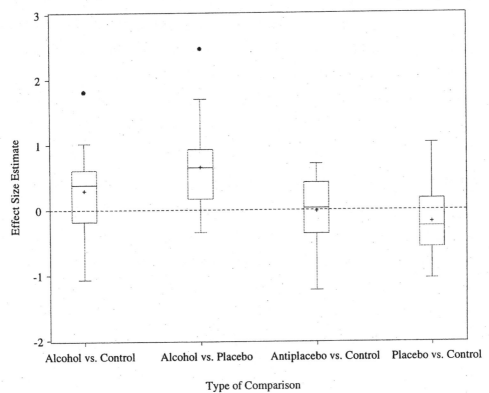

Bushman and Cooper (1990) viewed the set of side-by-side box plots as more informative than simply offering the reader an average effect-size estimate in each subgroup of studies. The average effect for the antiplacebo versus control comparison is about zero, suggesting that the pure pharmacological effects of alcohol do not increase aggression. The average effect for the placebo versus control comparison is also about zero, suggesting that the pure expectancy effects of alcohol do not increase aggression. When the pharmacological and expectancy effects of alcohol are combined, in the alcohol versus control and alcohol versus placebo conditions, the effect-size estimates are greater than zero. In the "real world," of course, the pharmacological and expectancy effects of alcohol are combined. The pattern of results depicted in Figure 3.15 suggests that the pharmacological and expectancy effects of alcohol interact to increase aggression.

Sometimes it is useful to see the box plot for all of the data together along with the box plots for each group separately, as shown in Figure 3.16. Duplicating each observation in the data set can produce this output. Thus, each observation appears once in the overall box plot and once in a separate box plot. The duplicate copy of each observation is given a constant value on the moderator variable, different from any of the actual values of the moderator variable. In the Bushman and Cooper (1990) example, the moderator variable for type of comparison is called TYPE, with value 1 for alcohol versus control comparison, value 2 for alcohol versus placebo comparison, value 3 for antiplacebo versus control comparison, and value 4 for placebo versus control comparison. We used the value TYPE=0 for the duplicate values. The following SAS code produces Figure 3.16:

```
/* side-by-side box plots;                  */
/*  including individual comparison types */
/*  and all four types combined            */
options nodate pagesize=54 linesize=80 nocenter pageno=1;
libname ch3 "d:\metabook\ch3\dataset";
data fig316;
   set ch3.bac90;
   output;           /* individual comparison types */
   type = 0;
   output;           /* all comparison types combined */
   label effect="Effect Size Estimate";
   label type="Type of Comparison";
proc sort data=fig316;
   by type;
filename fig316 "d:\tmp\fig316.cgm";
goptions reset=goptions device=cgmmwwc gsfname=fig316
   gsfmode=replace ftext=hwcgm005 gunit=pct;
```

```
proc shewhart data=fig316 gout=fig316 graphics;
   boxchart effect*type / boxstyle=schematic
                          idsymbol=dot
                          vref=0
                          haxis=axis1
                          nolegend
                          hoffset=5
                          nolimits
                          stddevs.;
axis1 value =
   ("All Types" "Alcohol vs. Control" "Alcohol vs. Placebo"
   "Antiplacebo vs. Control" "Placebo vs. Control");
run;
quit;
```

Figure 3.16 *Side-by-side box plots for all comparisons combined and for each type of comparison separately for the Bushman & Cooper meta-analysis*

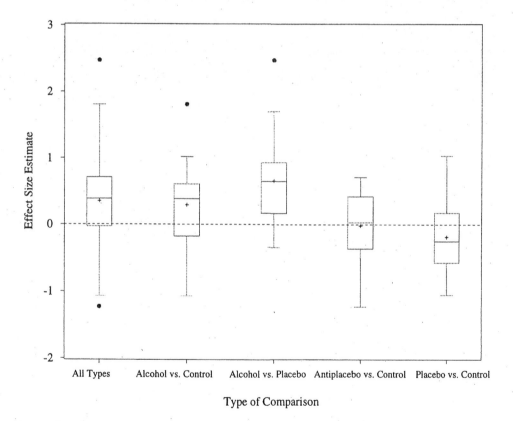

3.7 Conclusions

Simple graphic displays of effect-size estimates can greatly help the reader interpret the results of a meta-analysis, especially when the meta-analysis contains a large number of studies. Five types of plots were discussed in this chapter: dot plots, funnel plots, normal quantile plots, stem-and-leaf plots, and box plots.

Dot plots can be more useful than tables in depicting the individual study results in a meta-analysis. It is more useful to rank order studies by the effect-size estimate rather than by the first author's last name. It is also helpful to include a confidence interval with each effect-size estimate so that the reader can determine how reliable the estimate is and whether the estimate differs from zero.

You can use funnel plots and normal quantile plots to check whether the data come from a single population and whether publication bias is a problem. In addition, you can use normal quantile plots to check the normality assumption on which many meta-analytic procedures are based. We believe that normal quantile plots are more useful than funnel plots because they are easier to interpret. It is easier to determine if the study effects fall on a straight line and within 95% confidence interval bands than it is to determine if the study effects are shaped like a funnel.

Stem-and-leaf plots and box plots are especially efficient for presenting the major findings from a meta-analysis. They help the readers see the "big picture" without getting bogged down by details of the meta-analysis. The stem-and-leaf plot gives the shape of the distribution while retaining the actual effect-size estimates for the individual studies. No information is lost in a stem-and-leaf plot, and information is a meta-analyst's most precious commodity.

Box plots are especially useful in examining the symmetry of the distribution of effects and in identifying mild and severe outliers. When the magnitude of effect-size estimates is related to study characteristics, then graphs should encourage the reader to compare the estimates for different subgroups of the study characteristic (Light, Singer, & Willet, 1994). Side-by-side box plots are very useful for presenting the results for different subgroups of studies.

3.8 References

Bushman, B. J. & Cooper, H. M. (1990). Effects of alcohol on human aggression: An integrative research review. *Psychological Bulletin, 107,* 341–354.

Chambers, J. M., Cleveland, W. S., Kleiner, B., and Tukey, P. A. (1983). Graphical methods for data analysis. Monterey, CA: Wadsworth.

Cleveland, W. S. (1985). *The elements of graphing data.* Monterey, CA: Wadsworth.

Cohen, P. A. (1983). Comment on "A selected review of the validity of student ratings of teaching." *Journal of Higher Education, 54,* 449–458.

Devine, E. C. & Cook, T. D. (1983). A meta-analytic analysis of effects of psycho-educational interventions on length of hospital stay. *Nursing Research, 32,* 267–274.

Friendly, M. (1991). *SAS system for statistical graphics.* Cary, NC: SAS Institute Inc.

Greenwald, A. G. (1975). Consequences of prejudice against the null hypothesis. *Psychological Bulletin, 82,* 1–20.

Hedges, L. V. & Olkin, I. (1985). *Statistical methods for data analysis.* New York: Academic Press.

Kendall, M. G. & Stuart, A. (1977). *The advanced theory of statistics* (Vol. 1). Hafner, NY: John Wiley & Sons.

Light, R. J., & Pillemer, D. B. (1984). *Summing up: The science of reviewing research.* Cambridge, MA: Harvard University Press.

Light, R. J., Singer, J. D., & Willett, J. B. (1994). The visual presentation and interpretation of meta-analyses. In H. Cooper & L. V. Hedges (Eds.). *The handbook of research synthesis* (pp. 439–453). New York: Russell Sage Foundation.

Raudenbush, S. W. (1984). Magnitude of teacher expectancy effects on pupil IQ as a function of the credibility of expectancy induction: A synthesis of findings from 18 experiments. *Journal of Educational Psychology, 76,* 85–97.

Rosenthal, R. (1979). The "file-drawer problem" and tolerance for null results. *Psychological Bulletin, 86,* 638–641.

Ross, S., Krugman, A. D., Lyerly, S. B., & Clyde, D. J. (1962). Drugs and placebos: A model design. *Psychological Reports, 10,* 383–392.

SAS Institute Inc. (1990a). *SAS/GRAPH software: reference, version 6, 1st ed., volumes 1 & 2.* Cary, NC: SAS Institute Inc.

SAS Institute Inc. (1990b). *SAS procedures guide, version 6, 3rd ed.* Cary, NC: SAS Institute Inc.

SAS Institute Inc. (1994). TS-252E, *Exporting SAS/GRAPH Output to Microsoft PowerPoint for Windows.* Cary, NC: SAS Institute Inc.

SAS Institute Inc. (1995). *SAS/QC software: Usage and reference, version 6, 1st ed. volumes 1&2).* Cary, NC: SAS Institute Inc.

Tukey, J. W. (1977). *Exploratory data analysis.* Reading, MA: Addison-Wesley.

Wang, M. C., & Bushman, B. J. (1998). Using the normal quantile plot to explore meta-analytic data sets. *Psychological Methods, 3,* 46–54.

3.9 Appendices

Appendix 3.1 SAS Code for Output 3.1

```
libname ch3 'd:\metabook\ch3\dataset';
options nodate nocenter pagesize=54 linesize=132 pageno=1;
proc sort data=ch3.cohen83;
   by effect;
proc timeplot data=ch3.cohen83;
   plot lower="[" effect="*" upper="]" /
      overlay hiloc ref=0 refchar="0";
   id study;
   title;
run;
```

Appendix 3.2: SAS Macro for Finding the Minimum and Maximum Values of the Variables on the X and Y Axes

```
%macro findxy(dataname,xx,yy);
proc univariate data=&dataname noprint;
   var &xx &yy;
   output out=outa min = minxx minyy max = maxxx maxyy;
proc print data=outa;
run;
%mend findxy;
```

Appendix 3.3: SAS Macro for Entering Parameters for a Funnel Plot

```
%macro funnelin;
   %window FINDAT
      #5 @4 "Input name of SAS data set" @31 indat 15
         attr=underline
      #17 @29 "Press" @35 "ENTER" @41 "to continue.";
   %window FXVAR
      #5 @4 "Input name of variable on x-axis" @37 xvar 8
         attr=underline
      #17 @29 "Press" @35 "ENTER" @41 "to continue.";
   %window FXMIN
      #5 @4 "Input minimum value of X variable" @38 xmin 8
         attr=underline
      #17 @29 "Press" @35 "ENTER" @41 "to continue.";
   %window FXMAX
      #5 @4 "Input maximum value of X variable" @38 xmax 8
         attr=underline
      #17 @29 "Press" @35 "ENTER" @41 "to continue.";
   %window FXBY
      #5 @4 "Input value of BY in ORDER statement of X
variable"
         @55 xby 8 attr=underline
      #11 @4 "Example: Approximately equal to (XMAX-XMIN)/6"
      #17 @29 "Press" @35 "ENTER" @41 "to continue.";
   %window FYVAR
      #5 @4 "Input name of variable on y-axis" @37 YVAR 8
         attr=underline
      #17 @29 "Press" @35 "ENTER" @41 "to continue.";
   %window FYMIN
      #5 @4 "Input minimum value of Y variable" @38 YMIN
         8 attr=underline
      #17 @29 "Press" @35 "ENTER" @41 "to continue.";
   %window FYMAX
      #5 @4 "Input maximum value of Y variable" @38 YMAX 8
         attr=underline
      #17 @29 "Press" @35 "ENTER" @41 "to continue.";
   %window FYBY
      #5 @4 "Input value of BY in ORDER statement of Y
variable"
         @55 YBY 8 attr=underline
      #11 @4 "Example: Approximately equal to (YMAX-YMIN)/6"
      #17 @29 "Press" @35 "ENTER"  @41 "to continue.";
   %window FXLABEL
      #5 @4 "Input LABEL on x-axis:"
      #8 @4 XLAB 30 attr=underline
      #17 @29 "Press" @35 "ENTER" @41 "to continue.";
   %window FYLABEL
      #5 @4 "Input LABEL on y-axis:"
      #8 @4 YLAB 30 attr=underline
      #17 @29 "Press" @35 "ENTER" @41 "to continue.";
   %window FGDIR
      #5 @4 "Input directory name for Graph Out File"
```

```
        #8 @4 GDIR 30 attr=underline
        #17 @29 "Press" @35 "ENTER" @41 "to continue.";
    %window FGHOUT
        #5 @4 "Input name of graphical output file"
        #8 @4 GHOUT 30 attr=underline
        #17 @29 "Press" @35 "ENTER" @41 "to continue.";
    %display FINDAT;
    %display FXVAR;
    %display FXLABEL;
    %display FXMIN;
    %display FXMAX;
    %display FXBY;
    %display FYVAR;
    %display FYMIN;
    %display FYMAX;
    %display FYBY;
    %display FYLABEL;
    %display FGDIR;
    %display FGHOUT;
%mend funnelin;
```

Appendix 3.4: SAS Macro for Creating a Funnel Plot

```
%macro
funnel(indat,xvar,xlab,xmin,xmax,xby,yvar,ylab,ymin,ymax,yby
    ,gdir,ghout);
%funnelin;
filename &ghout "&gdir\&ghout..cgm";
goptions reset=goptions device=cgmmwwc gsfname=&ghout
gsfmode=replace
    ftext=hwcgm005 gunit=pct;
proc gplot data=&indat gout=ch3graph;
    PLOT &YVAR * &XVAR = 1 / VAXIS = AXIS1 VMINOR = 0
                            HAXIS = AXIS2 HMINOR = 0
                            NAME = "&GHOUT"
                            FRAME;
    symbol1 v=dot c=black height=1.5 i=none;
    symbol2 v=star c=black height=1.5 i=none;
    axis1 order=(&ymin to &ymax by &yby) value=(h=4)
        offset=(2) label=(h=4 a=90 r=0 &ylab) width=2;
    axis2 order=(&xmin to &xmax by &xby) length=80
        offset=(2) label=(h=4 &xlab) value=(h=4) width=2;
run;
quit;
%mend funnel;
```

Appendix 3.5: SAS Code for Figure 3.2

```
options nodate nocenter pagesize=54 linesize=80 pageno=1;
libname gdevice0 "c:\tmp";
/*******************************************************/
/*  FIG302: Funnel plot for simulation data set data302 */
/*******************************************************/
filename fig302 "c:\tmp\fig302.cgm";
/*******************************************************/
/* Data302: A simulated data set for Fig302            */
/* (1) For half of the samples in each study, the      */
/*     populations for the experimental                */
/*     and control groups are assumed to have normal   */
/*     distributions with mean 0 and variance 1        */
/* (2) For the other half of the samples in each study */
/*     the populations for the experimental and control*/
/*     groups are assumed to have normal distributions */
/*     with mean 0.8 and variance 1                    */
/* (3) Both groups have same sample sizes.             */
/* (4) Hedges' g effect-size estimate is used          */
/*******************************************************/
data temp;
   input group ne;
   seed=int(time());
   do study = 1 to 10;
      do i = 1 to ne;
         x=rannor(seed);
         output;
      end;
   end;
   keep group study ne x;
CARDS;
1       10
2       20
3       30
4       40
5       50
6       60
proc sort data=temp;
   by group study;
proc means data=temp noprint;
   var x ne;
   by group study;
   output out=tmpouta mean=mx ne std=stdx;
data temp;
   input group nc;
   seed=int(time());
   do study = 1 to 10;
      do i = 1 to nc;
         y=rannor(seed);
         output;
      end;
   end;
```

```
      keep group study nc y;
   CARDS;
   1        10
   2        20
   3        30
   4        40
   5        50
   6        60
   proc sort data=temp;
      by group study;
   proc means data=temp noprint;
      var y nc;
      by group study;
      output out=tmpoutb mean=my nc std=stdy;
   data data302a;
      merge tmpouta tmpoutb;
      effect = (mx-my)/sqrt(((ne-1)*stdx*stdx
               +(nc-1)*stdy*stdy)/(ne+nc-2));
      vareff = (ne+nc)/(ne*nc)+(effect*effect)/(2*(ne+nc-2));
      stdeff = effect / sqrt(vareff);
      sample = int(((sqrt(ne)+sqrt(nc))/2)
               *((sqrt(ne)+sqrt(nc))/2));
      pop=1;
      keep effect sample pop stdeff;
   data temp;
      input group ne;
      seed=int(time());
      do study = 1 to 10;
         do i = 1 to ne;
            x=rannor(seed)+0.8;
            output;
         end;
      end;
      keep group study ne x;
   CARDS;
   1        10
   2        20
   3        30
   4        40
   5        50
   6        60
   proc sort data=temp;
      by group study;
   proc means data=temp noprint;
      var x ne;
      by group study;
      output out=tmpouta mean=mx ne std=stdx;
   data temp;
      input group nc;
      seed=int(time());
      do study = 1 to 10;
         do i = 1 to nc;
            y=rannor(seed);
```

```
         output;
      end;
   end;
   keep group study nc y;
CARDS;
1        10
2        20
3        30
4        40
5        50
6        60
proc sort data=temp;
   by group study;
proc means data=temp noprint;
   var y nc;
   by group study;
   output out=tmpoutb mean=my nc std=stdy;
data data302b;
   merge tmpouta tmpoutb;
   effect = (mx-my)/sqrt(((ne-1)*stdx*stdx
            +(nc-1)*stdy*stdy)/(ne+nc-2));
   vareff = (ne+nc)/(ne*nc) + (effect*effect)/(2*(ne+nc-2));
   stdeff = effect / sqrt(vareff);
   sample = int(((sqrt(ne)+sqrt(nc))/2)
            *((sqrt(ne)+sqrt(nc))/2));

   pop=2;
   keep effect sample pop stdeff;
data data302;
   set data302a data302b;
run;
%findxy(data302,effect,sample);
goptions reset=goptions device=cgmmwwc gsfname=fig302
   gsfmode=replace ftext=hwcgm005 gunit=pct;
proc sort data=data302;
   by pop;
legend1 label=none frame down=1 across=2
        value=(h=4 "Population I" "Population II");
proc gplot data=ch3.data302 gout=ch3graph;
   plot sample * effect = pop
        / vaxis = axis1 vminor = 0
          haxis = axis2 hminor = 0
          name = 'fig302'
          legend = legend1
          frame;
   symbol1 v=dot c=black height=1.5 i=none;
   symbol2 v=circle c=black height=1.5 i=none;
   axis1 order=(10 to 60 by 10) offset=(2)
      label=(h=4 a=90 r=0 "Total Sample Size") value=(h=4)
      width=2;
   axis2 order=(-1.2 to 1.8 by 0.6) length=80 offset=(2)
      label=(h=4 "Effect Size Estimate") value=(h=4)
      width=2;
 run;
 quit;
```

Appendix 3.6: SAS Code Used to Create Normal Quantile Plots

```
%macro ciqqplot(indat,graphout,cgmfile,yvar);
proc sort data=&indat; by &yvar;
proc univariate noprint data=&indat;
   VAR &yvar;
   output out=stats n=nobs mean=mean std=std;
data quantile;
   set &indat;
   if _n_ = 1 then set stats;
   p = (_n_ - 0.5)/nobs;
   z = probit(p);
   normal = mean + z * std;
   se=(std/((1/sqrt(8*atan(1)))*exp(-0.5*z*z)))
      *sqrt(p*(1-p)/nobs);
   lower = normal - 2 * se;
   upper = normal + 2 * se;
   keep lower upper normal &yvar z;
proc gplot data=quantile gout=&graphout;
   plot &yvar * z = 1
        normal * z = 2
        lower  * z = 3
        upper  * z = 3
         / overlay
           vaxis = axis1 vminor = 0
           haxis = axis2 hminor = 0
           NAME = '&cgmfile'
           frame;
   symbol1 v=dot height=1.5 i=none color=black;
   symbol2 v=none i=joint l=3 color=black;
   symbol3 v=none i=joint l=20 color=black;
   axis1 offset=(2) label=(H=4 a=90 r=0 "Standardize Effect
      Size") value=(h=3) order = (-5 to 5 by 1.0) width=2;
   axis2 length=80 label=(h=4 "Normal Quantile")
       offset=(2) value=(h=4) width=2;
run;
quit;
%mend ciqqplot;
```

chapter 4

Combining Effect-Size Estimates Based on Categorical Data

4.1 Introduction

As was noted in Chapter 1, categorical data consist of frequency counts of observations that occur in different response categories. This chapter focuses on categorical variables that have only two levels (for example, smoker versus nonsmoker).

Section 4.2   describes two-way contingency tables.

Section 4.3   describes the odds ratio — the most commonly used measure of association for categorical data. The odds ratio has many practical and theoretical advantages over other measures of association for categorical data (for example, risk ratio or relative risk, difference between two probabilities, phi coefficient; see Agresti, 1990, pp. 16–18, and Fleiss, 1994, pp. 257–259).

Section 4.4   discusses how to combine odds ratios using the weighted average method.

Section 4.5   shows how to test whether odds ratios in a meta-analytical review are heterogeneous.

Section 4.6   discusses how to combine odds ratios using the Mantel-Haenszel method. The Mantel-Haenszel method is especially useful when several cells have small or zero frequency counts.

Section 4.7   describes how to control for the effects of covariates by using regression analysis and by using stratification.

Section 4.8   offers concluding comments.

4.2 Two-Way Contingency Tables

Let X and Y denote two variables, each having two levels or categories. In other words, variables X and Y are both binary response variables. When participants are classified on both variables, there are four (2 × 2) possible combinations of classifications. The responses (X, Y) of a participant who is randomly chosen from some population have a probability that can be displayed using a table that has two rows for the categories of X and two columns for the categories of Y. The cells of the table represent the four possible outcomes. Let π_{ij} denote the probability that (X, Y) falls in the cell in row i and column j. The probability distribution $\left\{\pi_{ij}\right\}$ is the joint distribution of X and Y. The marginal distributions are the row and column totals that are obtained by summing the joint probabilities. The marginal distributions are denoted by $\left\{\pi_{i+}\right\}$ for the row variable and by $\left\{\pi_{+j}\right\}$ for the column variable, where

$$\pi_{i+} = \sum_j \pi_{ij} \text{ and } \pi_{+j} = \sum_i \pi_{ij}, \text{ which satisfy } \sum_i \pi_{i+} = \sum_j \pi_{+j} = \sum_i \sum_j \pi_{ij} = 1. \text{ When}$$

the cells contain frequency counts of outcomes rather than probabilities, the table is called a contingency table. Table 4.1 gives the notation for joint, conditional, and marginal probabilities of a 2×2 table and Table 4.2 gives the notation for a 2×2 contingency table.

Table 4.1 *Notation for joint, conditional, and marginal probabilities of a 2×2 table*

Row (Variable X)	Column (Variable Y)		
	Positive	Negative	Total
Positive	π_{11}	π_{12}	π_{1+}
Negative	π_{21}	π_{22}	π_{2+}
Total	π_{+1}	π_{+2}	1.0

Table 4.2 *Notation for a 2 × 2 contingency table*

Row (Variable X)	Column (Variable Y)		
	Positive	Negative	Total
Positive	n_{11}	n_{12}	n_{1+}
Negative	n_{21}	n_{22}	n_{2+}
Total	n_{+1}	n_{+2}	n_{++}

The data in a contingency table can be generated by three major study designs: cross-sectional, prospective, and retrospective. In a cross-sectional design, participants report their status on variables X and Y at the same time. In a prospective design, participants report their status on variables X and Y at one time, and they report their status on variable Y at a subsequent time. In a retrospective design, participants report changes in their status on variable Y over some specified period of time, and they report their current status on variable X. Retrospective studies are distinguished from cross-sectional studies by the fact that participants report changes in their status on variable Y rather than their current status on variable Y.

For example, suppose that a researcher wants to know whether a vaccine (variable X) produces immunity to a certain flu strain (variable Y). In a cross-sectional design, the total number of participants n_{++} is fixed, and participants are classified according to whether they received the vaccine and whether they got the flu. In a prospective design, both the number of participants who received the vaccine n_{1+} and the number of participants who did not receive the vaccine n_{2+} are fixed, and participants are classified according to whether they subsequently got the flu. In a retrospective design, both the number of participants who got the flu n_{+1} and the number of participates who did not get flu n_{+2} are fixed, and participants are classified according to whether they had received the vaccine. Although the three study designs have different sampling distributions, the odds ratio is a valid measure of association in all three study designs.

4.3 The Odds Ratio ω

Consider a contingency table with underlying multinomial sampling distribution (in cross sectional designs) or product multinomial sampling distribution (in prospective or retrospective designs) given in Table 4.1. When the response for X is positive, the odds that the response for Y is positive is defined as

$$\Omega_1 = \frac{\pi_{11}}{\pi_{12}}. \tag{4.1}$$

When the response for X is negative, the odds that the response for Y is positive is defined as

$$\Omega_2 = \frac{\pi_{21}}{\pi_{22}}. \tag{4.2}$$

The ratio of the odds Ω_1 and Ω_2,

$$\omega = \frac{\Omega_1}{\Omega_2} = \frac{\pi_{11}/\pi_{12}}{\pi_{21}/\pi_{22}} = \frac{\pi_{11}\pi_{22}}{\pi_{12}\pi_{21}}, \tag{4.3}$$

is called the odds ratio. The odds ratio is also called the cross-product ratio because it is the ratio of the products of probabilities from diagonally opposite cells (that is, $\pi_{11}\pi_{22}$ and $\pi_{12}\pi_{21}$).

The odds ratio can range from 0 and ∞. When the value of the odds ratio equals 1, it means that variables X and Y are not associated. When the value of the odds ratio is greater than 1, it means that participants are more likely to receive a positive classification on variable Y if their classification on variable X is positive than if their classification on variable X is negative. For example, an odds ratio of 4 indicates that the odds of receiving a positive classification on variable Y is 4 times higher among participants who receive a positive classification on variable X than among participants who receive a negative classification on variable X.

The natural logarithm of the odds ratio, $\ln(\omega)$, is often used instead of the odds ratio for at least three reasons: (a) the log odds ratio converges more rapidly to a normal distribution than does the odds ratio (Agresti, 1990, p. 54), (b) the values for $\ln(\omega)$ and $\ln(1/\omega)$ represent the same level of association except for the sign (that is, $\ln(4) = 1.39$ and $\ln(1/4) = -1.39$), and (c) the value of the log odds ratio can range from $-\infty$ to ∞. When the value of the log odds ratio equals 0, it means that variables X and Y are not associated. When the value of the log odds ratio is greater than 0, it means that participants are more likely to receive a positive classification on variable Y if their classification on variable X is positive than if their classification on variable X is negative.

If the observed frequencies are displayed as in Table 4.2, the sample estimate of the odds ratio ω is defined as

$$\hat{\omega} = \frac{n_{11} n_{22}}{n_{12} n_{21}}. \tag{4.4}$$

The estimated large-sample variance of $\ln(\hat{\omega})$ is defined as

$$\mathrm{Var}\left\{\ln(\hat{\omega})\right\} = \left(\frac{1}{n_{11}} + \frac{1}{n_{12}} + \frac{1}{n_{21}} + \frac{1}{n_{22}} \right) \tag{4.5}$$

(Woolf, 1955). Once a $100(1-\alpha)\%$ confidence interval has been obtained for the log odds ratio, $\ln(\omega)$, the exponential function can be used to obtain a $100(1-\alpha)\%$ confidence interval for the odds ratio ω.

Because the sample odds ratio and the corresponding large-sample variance are not defined when one or more of the observed frequencies are equal to 0, it is good practice to add 0.5 to each observed frequency when one or more observed frequencies are small or zero (Gart & Zweiful, 1967; Haldane, 1955). Thus, the amended sample estimator of the odds ratio is defined as

$$\tilde{\omega} = \frac{(n_{11} + 0.5)(n_{22} + 0.5)}{(n_{12} + 0.5)(n_{21} + 0.5)}, \tag{4.6}$$

and the corresponding large-sample estimate of the variance for $\ln(\hat{\omega})$ is

$$\text{Var}\left\{\ln(\tilde{\omega})\right\} = \left(\frac{1}{n_{11}+0.5} + \frac{1}{n_{12}+0.5} + \frac{1}{n_{21}+0.5} + \frac{1}{n_{22}+0.5}\right). \tag{4.7}$$

Example 4.1 *Illustration of calculating estimates and 95% confidence intervals for the odds ratio, log odds ratio, amended odds ratio, and amended log odds ratio*

The data for this example come from a double-blind, placebo-controlled clinical trial of the effectiveness of the nicotine patch, a treatment for tobacco dependence, on smoking cessation (Imperial Cancer Research Fund General Practice Research Group, 1993; cited in Fiore et al., 1994, p. 1944). Participants in this clinical trial were randomly assigned to receive either a nicotine patch or a placebo patch for 12 weeks. At the end of treatment, researchers measured whether participants had quit smoking.

Table 4.3 *Contingency table for a double-blind, placebo-controlled clinical trial on the effectiveness of the nicotine patch on smoking cessation*

Row (X)	Column Y		
	Quit smoking	Didn't quit smoking	Total
Active patch	121	721	842
Placebo patch	73	771	844
Total	194	1,492	1,686

The SAS macro COMPODDS(*indata,outdata*) in Appendix 4.1 was used to compute the sample estimate and 95% confidence interval for the odds ratio, log odds ratio, amended odds ratio, and amended log odds ratio for the data in Table 4.3. *Indata* and *outdata* are the names for the input and output data sets, respectively. *Indata* contains four variables: N11, N12, N21, and N22, where NIJ is the frequency for cell I,J (see Table 4.2). *Outdata* contains four variables: type of effect-size (TYPE), effect-size estimate (ESTIMATE), and the lower and upper bounds of the 95% confidence interval for the effect size (LOWER and UPPER, respectively). The results from the following SAS code are printed in Output 4.1:

```
options nodate nocenter pagesize=54 linesize=80 pageno=1;
libname ch4 "d:\metabook\ch4\dataset";
data ch4.ex41;
    input n11 n12 n21 n22;
cards;
121 721 73 771
;
%compodds(ch4.ex41,ch4.ex41out)
proc format;
 value aa 1 = "Log odds ratio   "
          2 = "Odds ratio       "
          3 = "Amended log odds ratio"
          4 = "Amended odds ratio   ";
proc print data=ch4.ex41out noobs;
    title;
    var type estimate lower upper;
    format type aa. estimate lower upper 5.3;
run;
```

Output 4.1 *Illustration of calculating effect-size estimates for the odds ratio, log odds ratio, amended odds ratio, and amended log odds ratio*

TYPE	ESTIMATE	LOWER	UPPER
Log odds ratio	0.572	0.265	0.880
Odds ratio	1.772	1.303	2.411
Amended log odds ratio	0.570	0.263	0.877
Amended odds ratio	1.768	1.300	2.403

Because none of the cell frequencies in Table 4.3 are small or zero, you should use the odds ratio (or log odds ratio). As you can see in Output 4.1, the sample estimate for the odds ratio is 1.772 with a 95% confidence interval ranging from 1.303 to 2.411. Because the confidence interval does not include the value 1, more people quit smoking in the nicotine patch group than in the placebo patch group.

More specifically, people in the nicotine patch group were almost twice as likely to quit smoking as people in the placebo patch group.

4.4 Combining Odds Ratios Using the Weighted Average Method

Woolf (1955) proposed that a combined estimate of the log odds ratios be calculated simply as a weighted average of the log odds ratios in each study. This method assumes that the sample size for each of the k studies is large and that the odds ratios are homogeneous. The heterogeneity test for odds ratios is discussed in Section 4.5.

Let $\ln(\hat{\omega}_i)$ and $\text{Var}\{\ln(\hat{\omega}_i)\}$ be the sample log odds ratio and its corresponding large sample variance estimator in the ith study. In a fixed-effects meta-analysis, the weighted average estimator of the population log odds ratio $\ln(\omega)$ across the k studies is

$$\ln(\hat{\omega}_+) = \frac{\sum_{i=1}^{k} w_i \ln(\hat{\omega}_i)}{\sum_{i=1}^{k} w_i},$$
(4.8)

where the weight $w_i = 1/\text{Var}\{\ln(\hat{\omega}_i)\}$. Note that w_i gives studies with smaller variances (that is, larger sample sizes) more weight than studies with larger variances (that is, smaller sample sizes). Weighting effect-size estimates makes good theoretical sense because studies with smaller sample sizes produce less accurate effect-size estimates than studies with larger sample sizes.

The corresponding variance of $\ln(\hat{\omega}_+)$ is

$$\text{Var}\left\{\ln(\hat{\omega}_+)\right\} = \text{Var}\left(\frac{\sum\limits_{i=1}^{k} w_i \ln(\hat{\omega}_i)}{\sum\limits_{i=1}^{k} w_i}\right)$$

$$= \frac{\sum\limits_{i=1}^{k} w_i^2 \text{Var}\left\{\ln(\hat{\omega}_+)\right\}}{\left(\sum\limits_{i=1}^{k} w_i\right)^2} \quad . \tag{4.9}$$

$$= \frac{\sum\limits_{i=1}^{k} w_i^2 \left(\dfrac{1}{w_i}\right)}{\left(\sum\limits_{i=1}^{k} w_i\right)^2} = \frac{1}{\sum\limits_{i=1}^{k} w_i}$$

The $100(1-\alpha)\%$ confidence interval for $\ln(\omega)$ is given by

$$\ln(\hat{\omega}_+) - Z_{\alpha/2}\sqrt{\text{Var}\left\{\ln(\hat{\omega}_+)\right\}} \leq \ln(\omega) \leq \ln(\hat{\omega}_+) + Z_{\alpha/2}\sqrt{\text{Var}\left\{\ln(\hat{\omega}_+)\right\}}, \tag{4.10}$$

where $Z_{\alpha/2}$ is the two-sided critical value of the standard normal distribution at significance level α.

4.5 Heterogenity Test for Odds Ratios

Both the weighted average (Section 4.4) and the Mantel-Haenszel (Section 4.6) methods for estimating the common odds ratio are based on the assumption that the odds ratios are constant across studies. Because the heterogeneity test for odds ratios depends on the weighted average of the odds ratios, we postpone our discussion of the heterogeneity test until after our discussion of the weighted average method for combining odds ratios.

As with other effect-size measures, odds ratios should not be combined unless they are homogeneous. If any study has an odds ratio that is much larger or much smaller than the "average" odds ratio, then you would expect the estimated cell frequencies for this study to be significantly different from the estimated cell frequencies for the other studies. Thus, a reasonable chi-square test for the adequacy of the assumption of a common odds ratio is the sum of the squared deviations of observed and fitted cell frequencies, where each deviation is divided by its standard error (see Breslow & Day, 1980, pp. 142–148, for a more complete discussion of this test). Because there is no closed form equation for the fitted cell frequencies, they must be estimated numerically using a computer.

_____

Example 4.2 *Illustration of combining odds ratios using the weighted average method*

We are now ready to illustrate how to combine odds ratios and how to test whether the odds ratios in a meta-analysis are heterogeneous. The data for this example come from a meta-analysis of 17 double-blind, placebo-controlled clinical trials of the effectiveness of the nicotine patch on smoking cessation (Fiore el al., 1994). The clinical trial used in Example 4.1 was included in this meta-analysis. The results from all 17 clinical trials are given in Table 4.4.

Table 4.4 *Results from 17 double-blind, placebo-controlled clinical trials on the effectiveness of the nicotine patch on smoking cessation*

Clinical Trial	n_{11}	n_{12}	n_{21}	n_{22}
Abelin et al. (1989)	36	64	22	77
Abelin et al. (1989)	22	34	11	45
Buchremer et al. (1898)	29	13	22	21
Daughton et al. (1991)	19	36	7	45
Daughton et al. (1991)	20	31	7	45
Elan Pharmaceutical Research Corp.	45	110	22	142
Elan Pharmaceutical Research Corp.	48	91	30	107
Fiore et al. (1994)	21	36	11	44
Hurt et al., (1994)	56	64	24	96
Imperial Cancer Research Fund General Practice Research Group, (1993)	121	721	73	771
Mulligan et al., (1990)	19	21	6	34
Rusell et al., (1993)	70	330	15	185
Sachs et al. (1993)	46	67	17	90
Tonnesen et al., (1991)	43	102	7	137
Transdermal Nicotine Study Group, 1991	61	60	29	95
Transdermal Nicotine Study Group, 1991	37	91	11	118
Weisman et al., 1993	23	55	7	73

Note: n_{11} = number in the nicotine patch group who quit smoking. n_{12} = number in the nicotine patch group who didn't quit smoking. n_{21} = number in the placebo patch group who quit smoking. n_{22} = number in the placebo patch group who didn't quit smoking.

After computing an odds ratio for every study using the SAS macro in Appendix 4.1 if necessary, you can combine the odds ratios in Table 4.4 using the SAS macro WAVGODDS(*indata,outdata,intype,level*) in Appendix 4.2. *Indata* is the name of the output data set from the SAS macro COMPODDS. *Outdata* is the name of the output data set. *Intype* is a categorical variable that specifies whether you are combining log odds ratios (INTYPE=1) or amended log odds ratios (INTYPE=3). In this example, we used INTYPE=1 because none of the studies had small or zero cell frequencies. *Level* is the level of confidence for the confidence interval (for example, use 0.95 for a 95% confidence interval). The results from the following SAS code are printed in Output 4.2:

```
options nodate nocenter pagesize=54 linesize=80 pageno=1;
libname ch4 "d:\metabook\ch4\dataset";
%compodds(ch4.ex42,ch4.ex42out1);
%wavgodds(ch4.ex42out1,ch4.wavgodds,1,0.95);
proc format;
 value aa 1 = "Log Odds Ratio          "
          2 = "Odds Ratio              "
          3 = "Amended Log Odds ratio"
          4 = "Amended Odds Ratio      ";
proc print data=ch4.wavgodds noobs;
   title;
   format type aa. estimate level lower upper 5.3;
run;
```

Output 4.2 *Combined odds ratio estimate and 95% confidence interval for the meta-analysis on the effects of nicotine patches on smoking cessation*

TYPE	ESTIMATE	LEVEL	LOWER	UPPER
Log Odds Ratio	0.978	0.950	0.825	1.130
Odds Ratio	2.658	0.950	2.282	3.097

Output 4.2 includes both the natural logarithm of the combined odds ratio and its 95% confidence interval and the combined odds ratio and its 95% confidence interval. Because the confidence interval for the odds ratio does not include the value 1.0, the nicotine patch had a significant effect on smoking cessation. More specifically, participants in the nicotine patch group were 2.658 times more likely to quit smoking than were participants in the placebo patch group.

Example 4.3 *Heterogeneity test for odds ratios*

You can use the SAS macro HOMODDS (*indata,outdata,initial*) in Appendix 4.3, to compute the chi-square heterogeneity test statistic for the odds ratios in Table 4.4. The degrees of freedom for the chi-square statistic are $k-1$ (that is, the number of studies minus one). *Indata* and *outdata* are the names of the input and output data sets, respectively. *Indata* contains four variables: N11, N12, N21,

N22, where NIJ is the frequency for cell I,J (see Table 4.2). *Initial* is the common odds ratio estimate that is described in Section 4.4. The results from the following SAS code are printed in Output 4.3:

```
options nodate nocenter pagesize=54 linesize=80 pageno=1;
libname ch4 "d:\metabook\ch4\dataset";
%homodds(ch4.ex42,ch4.ex42out4,2,2.66);
proc print data=ch4.ex42out4 noobs;
   format df 3.0 chisq pvalue 5.3;
   title;
run;
```

Output 4.3 *Heterogeneity test of the odds ratios in the meta-analysis on the effectiveness of the nicotine patch on smoking cessation*

DF	CHISQ	PVALUE
16	25.08	0.068

Assume that you choose the $\alpha = .05$ significance level for testing the null hypothesis that the odds ratios are constant across studies. The chi-square statistic based on all 17 studies is 25.08 with 16 degrees of freedom, and the corresponding *p*-value for this statistic is 0.068. Because the *p*-value is greater than 0.5, it is reasonable to assume that the odds ratios are homogenous enough to combine.

4.6 Combining Odds Ratios Using the Mantel-Haenszel Method

The procedure for combining the odds ratios that are presented in Section 4.4 works well if the cell frequencies are large in each study and if the odds ratios are homogeneous. However, some studies might have small cell frequencies or even zero cell frequencies. Although you can use the amended odds or log odds ratio when some studies have small or zero frequencies, a better approach is to use a Mantel-Haenszel estimator (see Hauck, 1989). The Mantel-Haenszel procedure also assumes that the odds ratios are homogeneous. The common odds ratio estimator proposed by Mantel and Haenszel (1959) is

$$\hat{\omega}_{MH} = \frac{\sum_{i=1}^{k} n_{11_i} n_{22_i} / n_{++_i}}{\sum_{i=1}^{k} n_{12_i} n_{21_i} / n_{++_i}} . \tag{4.11}$$

The Mantel and Haenszel estimator is not affected by zero cell frequencies and it provides a consistent estimate of the common odds ratio even with several small cell frequencies.

Gart (1970) and Thomas (1975) provided an exact formula to compute a confidence interval for the common odds ratio estimate based on $\hat{\omega}_{MH}$. Because the exact formula requires extensive computation, however, we use the following large sample approximation estimate of the variance

$$\mathrm{Var}\left(\hat{\omega}_{MH}\right) = \frac{1}{2\left(\sum_{i=1}^{k} A_i\right)^2} \sum_{i=1}^{k} A_i B_i + \hat{\omega}_{MH} \sum_{i=1}^{k} \left(B_i C_i + A_i D_i\right) + \hat{\omega}_{MH}^2 \sum_{i=1}^{k} C_i D_i , \tag{4.12}$$

where $A_i = \dfrac{n_{11_i} n_{22_i}}{n_{++_i}}$, $B_i = \dfrac{n_{11_i} + n_{22_i}}{n_{++_i}}$, $C_i = \dfrac{n_{12_i} n_{21_i}}{n_{++_i}}$, and $D_i = \dfrac{n_{12_i} + n_{21_i}}{n_{++_i}}$.

(see Robins, Breslow, & Greenland, 1986, and Robins, Greenland, & Breslow, 1986).

Example 4.4 *Combining odds ratios using the Mantel-Haenszel method*

The data set for this example was taken from a meta-analysis of the relation between panic disorder and mitral valve prolapse syndrome (Katerndahl, 1993). According to the fourth edition of the *Diagnostic and Statistical Manual of Mental Disorders* (American Psychiatric Association, 1994), the essential feature of panic disorder is the presence of recurrent, unexpected panic attacks followed by at least one month of persistent concern about having other attacks, worry about the possible implications or consequences of the attacks, or a significant behavioral change related to the attacks. The panic attacks are not due to the direct physiological effects of a substance (for example, caffeine) or a medical condition, and they are not better accounted for by another mental disorder. A panic attack is defined as a discrete period of intense fear or discomfort that is accompanied by at least four of the following 13 symptoms: "palpitations, sweating, trembling or shaking, sensations of shortness of breath or smothering, feeling of choking, chest pain or discomfort, nausea or abdominal distress, dizziness or lightheadedness, derealization or depersonalization, fear of losing control or 'going crazy,' fear of dying, parenthesis, and chills or hot flushes" (American Psychiatric Association, 1994, p. 394).

The symptoms of panic attacks have been linked to a medical disorder called mitral valve prolapse syndrome. Mitral valve prolapse syndrome is a cardiovascular abnormality characterized by an excess of tissue of the mitral valve, causing the tissue to prolapse into the left atrium during systole (Pariser, Pinta, & Jones, 1978).

The data from this meta-analysis are given in Table 4.5. This is a good example for illustrating the Mantel-Haenszel estimator because several cells have small or zero frequencies.

Table 4.5 *Meta-analysis of the relation between panic attacks and mitral valve prolapse syndrome*

Study	Outcome Variable	n_{11}	n_{12}	n_{21}	n_{22}
Crowe et al. (1979)	MVP	8	12	2	18
Kantor et al. (1980)	MVP	8	17	2	21
Venkatesh et al. (1980)	MVP	5	16	1	19
Szmuilowicz & Flannery (1980)	Panic	13	18	44	82
Kane et al. (1981)	Panic	5	2	60	31
Uretsky (1982)	Panic	4	25	45	808
Hartman et al. (1982)	Panic	1	2	32	68
Pecorelli (1984)	Panic	2	3	0	25
Shear et al. (1984)	MVP	2	23	0	25
Chan et al. (1984)	MVP	0	15	0	4
Bass & Wade (1984)	MVP	0	17	0	29
Bowen et al. (1985)	Panic	0	0	16	14
Dunner (1985)	Panic	14	10	7	16
Nesse et al. (1985)	MVP	7	13	0	3
Dager et al. (1986)	MVP	8	16	3	17
Mazza et al. (1986)	Panic	0	0	48	49
Devereux et al. (1986)	Panic	8	5	73	167
Dager et al. (1987)	MVP	17	12	2	7
Wulsin et al. (1988)	Unspecified	0	8	2	39
Taylor (1988)	MVP	0	12	0	12
Gorman et al.	MVP	14	22	4	18

Note: Panic = panic disorder. MVP = mitral valve prolapse syndrome. n_{11} = number of cases with panic disorder coded positive and mitral valve prolapse syndrome coded positive. n_{12} = number of cases with panic disorder coded positive and mitral valve prolapse syndrome coded negative. n_{21} = number of cases with panic disorder coded negative and mitral valve prolapse syndrome coded positive. n_{22} = number of cases with panic disorder coded negative and mitral valve prolapse syndrome coded negative.

Because the Breslow-Day heterogeneity test statistic (see Section 4.5) was nonsignificant at the .05 level, $\chi^2(20, N = 21) = 13.33$, $p = .863$, it seems reasonable to include all 21 studies in the meta-analysis.

   The SAS macro MHODDS(*indata,outdata,level*) in Appendix 4.4 was used to compute the Mantel-Haenszel estimator and 95% confidence interval for the common odds ratio using the data in Table 4.5. *Indata* and *outdata* are the names of the input and output data sets, respectively. *Indata* contains four variables: N11, N12, N21, N22, where NIJ is the frequency for cell I,J (see Table 4.2). *Level* is the level of confidence for the confidence interval (for example, use 0.95 for a 95% confidence interval). The results from the following SAS code are printed in Output 4.4:

```
options nodate nocenter pagesize=54 linesize=80 pageno=1;
libname ch4 "d:\metabook\ch4\dataset";
%mhodds(ch4.ex44,ch4.mhodds,.95);
proc format;
    value bb 5="Log odds ratio (MH)"
             6="Odds ratio (MH)   ";
proc print data=ch4.mhodds noobs;
    var type estimate level lower upper;
    format type bb. estimate level lower upper 5.3;
run;
```

Output 4.4 Combining odds ratios using the Mantel-Haenszel method

TYPE	ESTIMATE	LEVEL	LOWER	UPPER
Odds ratio (MH)	2.830	0.950	2.457	3.203
Log odds ratio (MH)	1.040	0.950	0.899	1.164

As you can see in Output 4.4, the Mantel-Haenszel estimator of the common odds ratio is 2.830, with a 95% confidence interval ranging from 2.457 to 3.203. Katerndahl (1993) reported that the weighted average of the sample odds ratio was 2.3, with a 95% confidence interval ranging from 1.6 to 3.5. Because some studies have zero and small cell frequencies, the Mantel-Haenszel method yields a narrower confidence interval than does the weighted average method. The results from both procedures suggest that there is a significant relation between mitral valve prolapse syndrome and panic disorder.

4.7 Controlling for the Effects of Covariates

In Sections 4.4 and 4.6, we assumed that the contingency table from each study was observed under similar conditions (that is, without the presence of covariates). However, the contingency tables from different studies in a meta-analytical review are seldom observed under similar conditions. Thus, covariate effects should be controlled when combining odds ratios. We discuss two methods for controlling covariate effects: logistic regression and stratification.

4.7.1 Control by Logistic Regression

Let X and Y represent two binary response variables that are crossed to form a 2×2 contingency table, and let $Z_1,...,Z_p$ represent p covariates. You can use logistic regression analysis to determine the relation between variables X and Y after controlling for the relation between the covariates $Z_1,...,Z_p$ and variable Y (Agresti, 1990, page 85–90). In a meta-analysis, the covariates are study characteristics. (For a more detailed discussion about covariates in logistic regression analysis, see Agresti, 1997, pp. 575–619, and *Logistic Regression Examples Using the SAS System*, 1995.)

If the values 1 and 0 are used for the two categories of variable Y, then Y is a Bernoulli random variable with mean $E(Y) = \Pr(Y = 1) = \Pi$. If the logistic regression model only includes the explanatory variable X, the conditional mean of Y is $E(Y \mid X) = \Pr(Y = 1 \mid X) = \Pi$. If the logistic regression model includes the explanatory variable X and the p covariates $Z_1,...,Z_p$, the conditional mean of Y is $E(Y \mid X, Z_1,...,Z_p) = \Pr(Y = 1 \mid X, Z_1,...,Z_p) = \Pi$. This conditional probability is given by

$$\Pi = \frac{e^{\beta_0+\beta_1 X+\alpha_1 Z_1+\cdots+\alpha_p Z_p}}{1 + e^{\beta_0+\beta_1 X+\alpha_1 Z_1+\cdots+\alpha_p Z_p}}. \tag{4.13}$$

The CATMOD (CATegorical data analysis MODEL) procedure can be used to obtain the predicted cell frequencies for each study after controlling for the influence of covariates. After the predicted cell frequencies have been obtained for each study, you can use the SAS macro COMPODDS in Appendix 4.1 to obtain an odds ratio for each study. The odds ratios can then be combined using the SAS macro MHODDS in Appendix 4.4 if some of the studies have small or zero predicted frequencies, or using the SAS macro WAVGODDS in Appendix 4.2 otherwise.

Example 4.5 *Combining odds ratios after controlling for covariates using logistic regression analysis*

Table 4.6 contains the covariates for each study in the meta-analysis of double-blind, placebo-controlled clinical trials on the effectiveness of the nicotine patch on smoking cessation (Fiore et al., 1994; see Example 4.2).

Table 4.6 *Covariates examined in the meta-analysis of 17 double-blind, placebo-controlled clinical trials on the effectiveness of the nicotine patch on smoking cessation*

Clinical trials	Patch type, h	Weeks of treatment	Weeks of Weaning	Female, %	Mean age, y	Mean no. cigarettes/d
Abelin et al. (1989)	24	12	8	40.2	41.6	27.5
Abelin et al. (1989)	24	9	6	26.8	24.0	23.0
Buchremer et al. (1898)	24	7	1	50.4	35.3	29.3
Daughton et al. (1991)	16	4	0	53.2	41.8	32.9
Daughton et al. (1991)	24	4	0	53.2	41.8	32.9
Elan Pharmaceutical Research Corp.	24	6	0	58.4	41.3	30.8
Elan Pharmaceutical Research Corp.	24	8	0	64.5	41.3	29.3
Fiore et al. (1994)	24	6	2	67.9	43.6	30.3
Hurt et al., (1994)	24	8	0	53.8	43.2	28.6
Imperial Cancer Research Fund General Practice Research Group, (1993)	24	12	8	55.1	43.7	24.4
Mulligan et al., (1990)	24	6	0	52.5	36.9	25.1
Rusell et al., (1993)	16	18	6	61.2	39.5	23.5
Sachs et al. (1993)	16	18	6	59.1	47.6	28.0
Tonnesen et al., (1991)	16	16	4	70.0	45.2	21.5
Transdermal Nicotine Study Group, 1991	24	12	6	60.3	43.3	31.2
Transdermal Nicotine Study Group, 1991	24	12	6	62.7	43.1	30.5
Weisman et al., 1993	24	6	2	56.6	41.8	41.8

Note: h = hours, y = years, d = day.

The SAS code in Appendix 4.5 was used to obtain the predict cell frequencies using logistic regression method. In this example, the response variable was smoking cessation (1 = quit smoking, 0 = didn't quit smoking). The explanatory variable was treatment condition (that is, nicotine versus placebo patch). The covariates included (a) daily patch duration (that is, 16-hour versus 24-hour), (b) patch treatment duration, (c) weaning or dosage-reduction duration, (d) percent of participants who were females, (e) age of participants, and (f) mean number of cigarettes smoked per day. Because all of the covariates are continuous variables, they appear in both the direct and the model statements. Because the explanatory variable is categorical, it only appears in the MODEL statement.[1]

Output 4.5 contains the output data set of the fitted cell frequencies that are obtained from the SAS code in Appendix 4.5.

Output 4.5 *Cell frequencies after control by covariates for 17 double-blind, placebo-controlled clinical trials on the effectiveness of the nicotine patch on smoking cessation*

STUDYID	PN11	PN12	PN21	PN22
Abelin et al. (1989)	30.8	69.2	17.7	81.3
Abelin et al. (1989)	18.6	37.4	11.3	44.7
Buchremer et al. (1898)	14.9	27.1	9.6	33.4
Daughton et al. (1991)	12.8	42.2	5.3	46.7
Daughton et al. (1991)	17.5	33.5	11.1	40.9
Elan Pharmaceutical Research Corp.	52.2	102.8	33.9	130.1
Elan Pharmaceutical Research Corp.	45.7	93.3	27.2	109.8
Fiore et al. (1994)	11.9	45.1	4.3	50.7
Hurt et al., (1994)	49.0	71.0	33.4	86.6
Imperial Cancer Research Fund (1993)	166.6	675.4	57.5	786.5
Mulligan et al., (1990)	13.7	26.3	8.5	31.5
Rusell et al., (1993)	86.4	313.6	17.2	182.8
Sachs et al. (1993)	31.2	81.8	15.7	91.3
Tonnesen et al., (1991)	27.8	117.2	8.9	135.1
Transdermal Nicotine Study Group,1991	30.6	90.4	15.3	108.7
Transdermal Nicotine Study Group 1991	29.8	98.2	13.3	115.7
Weisman et al., 1993	25.5	52.5	15.8	64.2

---

[1] If you are interested in learning more about PROC CATMOD, consult Chapter 17 in *SAS/STAT USER's Guide* (1990) and *Logistic Regression Examples Using the SAS System* (1995).

The SAS macro COMPODDS (Appendix 4.1) was used to obtain the odds ratio for each study in Output 4.5. Because none of the cell frequencies are small or zero, the SAS macro WAVGODDS (Appendix 4.2) was used to obtain an estimate and 95% confidence interval for the true odds ratio. The results are depicted in Output 4.6.

Output 4.6 *Combined odds ratio estimate and 95% confidence interval for the meta-analysis on the effects of nicotine patches on smoking cessation after controlling for covariates*

TYPE	ESTIMATE	LEVEL	LOWER	UPPER
Log Odds Ratio	0.893	0.950	0.740	1.047
Odds Ratio	2.443	0.950	2.095	2.849

Output 4.6 contains both the natural logarithm of the combined odds ratio and its 95% confidence interval and the combined odds ratio and its 95% confidence interval. Because the confidence interval for the odds ratio does not include the value 1.0, the nicotine patch had a significant effect on smoking cessation. Participants in the nicotine patch group were 2.443 times more likely to quit smoking than were participants in the placebo patch group. Note that the confidence interval is narrower when covariates are controlled (2.849 − 2.095 = 0.754) than when covariates are not controlled (3.097 − 2.282 = 0.815).

The corresponding chi-square test for heterogeneity of odds ratios is given in Output 4.7.

Output 4.7 *Heterogeneity test of the odds ratios in the meta-analysis on the effectiveness of the nicotine patch on smoking cessation after controlling for covariates*

```
DF   CHISQ   PVALUE

16   12.36   0.719
```

Assume that you choose the $\alpha = .05$ significance level for testing the null hypothesis that the odds ratios are constant across studies. The chi-square statistic based on all 17 studies is 12.36 with 16 degrees of freedom, and the corresponding *p*-value for this statistic is 0.719. Because the *p*-value is greater than .05, it is reasonable to assume that the odds ratios are homogeneous enough to combine. Note that the odds ratios are more homogeneous when covariates are controlled than when covariates are not controlled (compare Output 4.3 with Output 4.7).

4.7.2 Control by Stratification

Although you can use the logistic regression method that is discussed in Section 4.7.1 to control for covariates that are categorical study characteristics, an alternative approach is to use stratification. First, the contingency tables are classified into different strata based on the covariates such that the contingency tables in the same stratum are as similar as possible. Because the contingency tables in the same stratum are similar, a common odds ratio can be obtained for each stratum. Let $\hat{\omega}_{MH_s}$ and $\mathrm{Var}(\hat{\omega}_{MH})_s$ be the Mantel-Haenzsel estimator and corresponding variance estimator for the log odds ratio in stratum s. A weighted average of the log odds ratios from each stratum is given by

$$O = \frac{\sum_{s=1}^{k} \mathrm{Var}(\hat{\omega}_{MH})_s \times \hat{\omega}_{MH_s}}{\sum_{s=1}^{k} \mathrm{Var}(\hat{\omega}_{MH})_s}. \tag{4.14}$$

The estimated variance of O is given by :

$$\text{Var}(O) = \frac{1}{\sum_{s=1}^{k} \text{Var}(\hat{\omega}_{MH})_s}. \tag{4.15}$$

Example 4.6 *Combining odds ratios after controlling for covariates using stratification*

This example illustrates how to control for covariates using stratification. In the meta-analysis on the relation between panic disorder and mitral valve prolapse syndrome (Katerndahl, 1993, see Example 4.4), studies differed according to whether the outcome variable was panic disorder or mitral valve prolapse syndrome. All studies in the meta-analysis used cross-sectional designs. In 11 studies, researchers measured the prevalence of mitral valve prolapse syndrome in panic patients. In nine studies, researchers measured the prevalence of panic disorder in mitral valve prolapse patients. In one study, no information is given about what patients were being treated for. The Breslow-Day heterogeneity test (see Section 4.5) was nonsignificant for both MVP and Panic outcome variables, $\chi^2(10,\ N=11) = 3.466,\ p=.97$ and $\chi^2(8,\ N=9)=8.563,\ p=.38$, respectively.

The SAS macro MHODDS in Appendix 4.4 was used to obtain the common log odds ratio. Output 4.8 (outcome variable = MVP) and Output 4.9 (outcome variable = Panic) were produced using the following SAS code:

```
options nodate nocenter pagesize=54 linesize=80 pageno=1;
libname ch4 "d:\metabook\ch4\dataset";
data temp;
   set ch4.ex44;
   if (outcome = "MVP    ");
%mhodds(temp,ch4.mvpodds,.95);
proc format;
   value bb 5="Log odds ratio (MH)"
            6="Odds ratio (MH)  ";
proc print data=ch4.mvpodds noobs;
   var type estimate level lower upper;
   format type bb. estimate level lower upper 6.3;
run;
data temp;
   set ch4.ex44;
   if (outcome = "Panic   ");
%mhodds(temp,ch4.panodds,.95);
proc print data=ch4.panodds noobs;
   var type estimate level lower upper;
   format type bb. estimate level lower upper 5.3;
run;
```

Output 4.8 *Combining odds ratios using the Mantel-Haenszel method for the prevalence of mitral valve prolapse syndrome in panic patients*

TYPE	ESTIMATE	LEVEL	LOWER	UPPER
Odds ratio (MH)	4.434	0.950	3.783	5.085
Log odds ratio (MH)	1.489	0.950	1.331	1.626

Output 4.9 *Combining odds ratios using the Mantel-Haenszel method for the prevalence of panic disorder in mitral valve prolapse patients*

TYPE	ESTIMATE	LEVEL	LOWER	UPPER
Odds ratio (MH)	2.199	0.950	1.729	2.670
Log odds ratio (MH)	0.788	0.950	0.547	0.982

Because the odds ratios for the two outcome variables were significantly different ($Z=5.45$, $p < .05$), the odds ratios should not be combined. The Mantel-Haenszel odds ratio for the two outcome variables should be reported separately. Recall that the Mantel-Haenszel common odds ratio in Example 4.4 was 2.83, a value between 4.434 for the MVP outcome variable and 1.899 for the Panic outcome variable.

4.8 Conclusion

The odds ratio is the effect-size index of choice for categorical data because it has some nice statistical properties. Before odds ratios from different studies are combined, the meta-analyst should test the assumption that the odds ratios are constant across studies. When the observed cell frequencies are small, 0.5 can be added to each cell frequency to obtain the amended odds ratio. A better approach, however, is to use the Mantel-Haenszel estimator.

Methods for combining odds ratios assume that the contingency table from each study was observed under similar conditions, that is, without the presence of covariates. However, the contingency tables from different studies in a meta-analysis are seldom observed under similar conditions. The meta-analyst should, therefore, control for the presence of covariates before combining the odds ratios. This chapter discussed two methods for controlling the effect of covariates: logistic regression and stratification. In Example 4.5, the odds ratios were more homogeneous and the confidence interval for the common odds ratio was narrower when covariates were controlled using logistic regression analysis than when covariates were ignored. In Example 4.4, the chi-square heterogeneity test was nonsignificant, indicating that the variation in odds ratios across studies was not greater than would be expected by chance alone. In Example 4.6, however, closer examination of the odds ratios in different strata revealed that the odds ratios from the two strata should not be combined. These examples illustrate the importance of controlling for covariates when combining odds ratios.

4.9 References

Agresti, A. (1990). *Categorical data analysis*. New York: John Wiley & Sons.

Agresti, A. (1997). *Statistical methods for the social sciences* (3rd ed.). New Jersey: Prentice Hall.

American Psychiatric Association (1994). *Diagnostic and statistical manual of mental disorders* (4th ed.) Washington, DC: Author.

Breslow, N. E. & Day, N. E. (1980). *Statistical methods in cancer research*. Lyon, France: Internal Agency for Research in Cancer.

Fiore, M. C., Smith, S. S., Jorenby, D. E., & Baker, T. B. (1994). The effectiveness of the nicotine patch for smoking cessation. *Journal of the American Medical Association, 271*, 1940–1947.

Fleiss, J. L. (1994). Measures of effect size for categorical data. In H. Cooper & L. V. Hedges (Eds.), *The handbook of research synthesis* (pp. 245–260). New York: Russell Sage Foundation.

Gart, J. J. (1970). Point and interval estimation of the common odds ratio in the combination of 2 × 2 tables with fixed margins. *Biometrika, 57*, 471–475.

Gart, J. J. & Zweiful, J. R. (1967). On the bias of various estimators of the logit and its variance with applications to quantal bioassay, *Biometrika, 54*, 181–187.

Haldane, J. B. S. (1955). The estimation and significance of the logarithm of a ratio of frequencies. *Annals of Human Genetics, 20*, 309–311.

Hauck, W. W. (1989). Odds ratio inference from stratified samples. *Communications in Statistics, 18A*, 767–800.

Imperial Cancer Research Fund General Practice Research Group (1993). Effectiveness of a nicotine patch in helping people stop smoking: Results of a randomized trial in a general practice. *British Medical Journal, 306*, 1304–1308 .

Katerndahl, D. A. (1993). Panic and prolapse meta-analysis. *Journal of Nervous and Mental Diseases, 181*, 539–544.

Mantel, N. & Haenszel, W. (1959). Statistical aspects of the analysis of data from retrospective studies of disease. *Journal of National Cancer Research, 22*, 719–748.

Pariser, S. F., Pinta, E. R., & Jones, B. A. (1978). Mitral valve prolapse syndrome and anxiety neurosis/panic disorder. *American Journal of Psychiatry, 135*, 246–247.

Robins, J., Breslow, N. E., & Greenland, S. (1986). Estimators of the Mantel-Haenszel variance consistent in both sparse data and large-strata limiting models. *Biometrics, 42*, 311–323.

Robins, J., Greenland, S., & Breslow, N. E. (1986). A general estimator for the variance of the Mantel-Haenszel odds ratio. *American Journal of Epidemiology, 124*, 719–723.

SAS Institute Inc. (1995). *Logistic regression examples using the SAS system, version 6, 1st ed.*, Cary, NC: Author.

SAS Institute Inc. (1990). *SAS/STAT user's guide, version 6, 4th ed., vol. 1 and 2*. Cary, NC: Author.

Thomas, D. G. (1975). Exact and asymptotic methods for the combination of 2×2 tables, *Computation Biomedical Research, 8,* 423–446.

Woolf, B. (1955). On estimating the relation between blood group and disease. *Annals of Human Genetics, 19,* 251–253.

4.10 Appendices

Appendix 4.1: SAS Macro for Computing the Odds Ratio

```
%macro compodds(indata,outdata);
/**********************************************************/
/* INDATA:                                             */
/*   N11: Sample frequency for cell (1,1)              */
/*   N12: Sample frequency for cell (1,2)              */
/*   N21: Sample frequency for cell (2,1)              */
/*   N22: Sample frequency for cell (2,2)              */
/*                                                     */
/* OUTDATA:                                            */
/*   TYPE:                                             */
/*     TYPE=1:    Log odds ratio                       */
/*     TYPE=2:    Odds ratio                           */
/*     TYPE=3:    Amended log odds ratio               */
/*     TYPE=4:    Amended odds ratio                   */
/*   ESTIMATE: Sample estimate for given TYPE          */
/*   LOWER: Lower bound for 95% confidence interval    */
/*   UPPER: Upper bound for 95% confidence interval    */
/**********************************************************/
data &outdata; set &indata;
  if (n11 = 0 or n12 = 0 or n21 = 0 or n22 = 0) then do;
   odds = .;
   logodds = .;
   variance = .;
   std = .;
   end;
  else do;
   odds = (n11*n22)/(n12*n21);
   variance = 1/n11+1/n12+1/n21+1/n22;
   logodds = log(odds);
   std = sqrt(variance);
  end;
  type = 1;
  estimate = logodds;
  lower = logodds + probit(0.025) * std;
  upper = logodds + probit(0.975) * std;
  output;
  type = 2;
  estimate = odds;
  lower = exp(lower);
  upper = exp(upper);
  variance = .;
  output;
  odds = (n11+.5)*(n22+.5)/((n12+.5)*(n21+.5));
  variance = 1/(n11+.5)+1/(n12+.5)+1/(n21+.5)+1/(n22+.5);
  std = sqrt(variance);
  type = 3;
  estimate = log(odds);
  lower = estimate + probit(.025)*std;
```

```
upper = estimate + probit(.975)*std;
output;
type = 4;
estimate = odds;
lower = exp(lower);
upper = exp(upper);
variance = .;
output;
keep type estimate lower upper variance;
%mend compodds;
```

Appendix 4.2: SAS Macro for Computing the Common Odds Ratio Based on the Weighted Average Method

```
%macro wavgodds(indata,outdata,intype,level);
/********************************************************/
/* INDATA: The output data set from SAS MACRO COMPODDS  */
/* INTYPE: Type of log odds ratio to be combined        */
/*   INTYPE = 1 for combining log odds ratio            */
/*   INTYPE = 3 for combining amended log               */
/*       odds ratio because some studies have zero      */
/*       frequencies in one or more cells               */
/* LEVEL: Level of confidence for the confidence        */
/*          interval of the combined weighted average   */
/*                                                      */
/* OUTDATA:                                             */
/*   The output data file contains five variables.      */
/*                                                      */
/*   (1) TYPE: The type of effect size                  */
/*       TYPE can have value 1 (log odds ratio) and     */
/*       2 (odds ratio) if INTYPE = 1.                  */
/*       TYPE can have value 3 (amended log odds        */
/*       ratio) and 4 (amended odds ratio) if           */
/*       INTYPE = 3.                                    */
/*   (2) ESTIMATE: Combined estimate for the            */
/*       given effect size TYPE.                        */
/*   (3) LOWER: Lower bound for confidence interval     */
/*   (4) UPPER: Upper bound for confidence interval     */
/*   (5) LEVEL: Level of confidence for the             */
/*       confidence interval.                           */
/********************************************************/
data;
  set &indata;
  if type = &intype;
  weight = 1 / variance;
  wlodds = estimate * weight;
  keep wlodds weight;
proc means noprint;
  var wlodds weight;
  output out = &outdata sum=swlodds sweight;
data &outdata;
  set &outdata;
  type = &intype;
  estimate = swlodds / sweight;
  level=&level;
  variance = 1 / sweight;
  lower = estimate + probit(.5-.5*&level) * sqrt(variance);
  upper = estimate + probit(.5+.5*&level) * sqrt(variance);
  output;
  type = &intype+1;
  estimate = exp(estimate);
  lower = exp(lower);
  upper = exp(upper);
  output;
  keep type estimate level lower upper;
%mend wavgodds;
```

Appendix 4.3: SAS Macro for Heterogeneity Test of Odds Ratios

```
options nodate nocenter pagesize=54 linesize=80 pageno=1;
%macro homeodds(indata,outdata,type,initial);
data; set &indata;
/*****************************************************/
/* Compute the chi-square test statistics for the    */
/* homogeneity of common odds ratio based on Breslow  */
/* and Day (1980)                                     */
/*                                                    */
/* INPUT DATA:                                        */
/*  Cell frequencies for each study:                  */
/*    N11, N12, N21, and N22.                          */
/*                                                    */
/* OUTPUT DATA:                                       */
/*  The chi-square test statistics with k-1 degrees of */
/*  freedom, where k is the number of studies in the   */
/*  meta-analysis.                                     */
/*                                                    */
/*                                                    */
/* TYPE:                                              */
/*  TYPE=2 if no studies have zero cell frequencies    */
/*  TYPE=4 if some studies have zero cell frequencies  */
/*                                                    */
/* INITIAL:                                           */
/*  Initial value to start the Newton iteration. The   */
/*  weighted average common odds ratio can be used as  */
/*  the initial value.                                 */
/*****************************************************/
  if (&type=4) then do;
     n11=n11+.5;
     n12=n12+.5;
     n21=n21+.5;
     n22=n22+.5;
  end;
  oddmle=&initial;
  df = 1;
  m1=n11+n22;
  m2=n12+n21;
  xold=n11;
  xnew=(n11+n12)*(n11+n21)/(n11+n12+n21+n22);
/*****************************************************/
/*  Estimate the cell frequencies based on the Newton */
/*  method                                    */
/*****************************************************/
  do until (abs(xold-xnew) < .1e-5);
     xold = xnew;
     fold = ((1-oddmle)*xnew-(m1+oddmle*m2))*xnew
            +(n11*n22-n12*n21*oddmle);
     pfold = 2*(1-oddmle)*xnew-(m1+oddmle*m2);
     xnew = xold - fold / pfold;
  end;
```

```
        varstat=1/(1/(n11-xnew)+1/(n12+xnew)+1/(n21+xnew)+1/
        (n22-xnew));
        chisq = (xnew*xnew)/varstat;
        keep df chisq;
proc means noprint;
        var df chisq;
        output out=&outdata sum = df chisq;
data &outdata;
        set &outdata;
        df = df -1;
        pvalue = 1-probchi(chisq,df);
        keep df chisq pvalue;
run;
%mend homeodds;
```

Appendix 4.4: SAS Macro for Computing the Common Odds Ratio Based on the Mantel-Haenszel Method

```
%macro mhodds(indata,outdata,level);
/****************************************************************/
/* Compute the odds ratio for each study                       */
/* (1) Compute Mantel-Haenszel odds ratio and log odds         */
/*     ratio                                                     */
/* (2) Compute the large sample variance estimator based*/
/*     on Robins, Breslow, and Greenland (1986) and           */
/*     Robins, Greenland, and Breslow (1986)                   */
/*                                                              */
/* INPUT DATA:                                                  */
/*   N11, N12, N21, N22: Cell frequencies for each             */
/*   study.                                                     */
/*   LEVEL: Level of confidence for the confidence             */
/*   interval                                                   */
/*                                                              */
/* OUTPUT DATA:                                                 */
/*   TYPE:  5 = Log of Mantel-Haneszel odds ratio              */
/*              estimate                                         */
/*          6 = Mantel-Haneszel odds ratio estimate            */
/*                                                              */
/*   ESTIMATE: Mantel-Haneszel estimate for the given          */
/*             type                                             */
/*   LOWER:  Lower bound of confidence interval                */
/*   UPPER:  Upper bound of confidence interval                */
/****************************************************************/
data;
   set &indata;
   ntt=n11+n12+n21+n22;
   aa=n11*n22/ntt;
   bb=(n11+n22)/ntt;
   cc=n12*n21/ntt;
   dd=(n12+n21)/ntt;
   ab=aa*bb;
   bc=bb*cc;
   ad=aa*dd;
   cd=cc*dd;
proc means noprint;
   var aa bb cc dd ab bc ad cd;
   output out=&outdata
          sum=saa sbb scc sdd sab sbc sad scd;
data &outdata;
   set &outdata;
   type=6;
   estimate=saa/scc;
   level=&level;
   variance=(sab+(saa/scc)*(sbc+sad+(saa/scc)*scd))
            /(2*saa*saa);
```

```
   lower=estimate+probit(.5-.5*level)*sqrt(variance);
   upper=estimate+probit(.5+.5*level)*sqrt(variance);
   output;
   type=5;
   estimate=log(saa/scc);
   lower=log(lower);
   upper=log(upper);
   variance=.;
   output;
   keep type estimate level variance lower upper;
%mend mhodds;
```

Appendix 4.5: SAS Code for Example 4.5

```
options nodate nocenter pagesize=54 linesize=80 pageno=1;
libname ch4 "d:\metabook\ch4\dataset";
proc catmod data=ch4.ex45b;
    weight count;
    direct ptype wtrt wwear female meanage meanciga;
    response joint / out=pred1;
    model quit = nicotine ptype wtrt wwear female meanage
                            meanciga / noparm noresponse pred=freq;
data temp; set pred1;
    keep nicotine _obs_ _pred_ _sample_ _number_;
data a1; set temp;
    if nicotine="yes";
    prob1 = _pred_;
    keep prob1;
data a2; set temp;
    if nicotine="no ";
    prob2 = _pred_;
    keep _sample_ prob2;
data temp;
    merge a2 a1;
data ch4.ex45out1;
    merge ch4.ex45a temp;
    total=n11+n12;
    pn11=total*(1-prob1);
    pn12=total-pn11;
    total=n21+n22;
    pn21=total*(1-prob2);
    pn22=total-pn21;
    keep studyid id pn11 pn12 pn21 pn22;
proc sort data=ch4.ex45out1;
    by id;
proc print data=ch4.ex45out1 noobs;
    var studyid pn11 pn12 pn21 pn22;
    format pn11 pn12 pn21 pn22 5.1;
    title;
run;
```

chapter 5

Combining Effect-Size Estimates Based on Continuous Data

5.1 Introduction

Chapter 4 described procedures for combining effect-size estimates based on categorical data. This chapter describes procedures for combining effect-size estimates based on continuous data. We begin by distinguishing between two families of effect-sizes: the standardized mean difference family and the correlation family.

Section 5.2.1	presents several estimators of the population standardized mean difference, including Glass's (1976) $\hat{\Delta}$, Cohen's (1969) d, Hedges' (1981) g, and Hedges' (1981) g_U. Because it has nice statistical properties, we recommend the use of Hedges unbiased estimator g_U.
Section 5.2.2	presents the Pearson product-moment correlation coefficient as an estimator of the population correlation coefficient.
Section 5.2.3	describes the point-biserial correlation to show the relation between the standardized mean difference and the correlation coefficient.
Section 5.3	describes how to convert test statistics to effect-size estimates and how to convert effect-size estimators from one type to another.
Sections 5.4 and 5.5	discuss how to obtain a weighted average estimate for the population standardized mean difference and the population correlation coefficient, respectively. To simplify our discussion, we assume that the effect-size estimates are homogeneous. Chapters 8 and 9 discuss how to formally test the homogeneity assumption.

5.2 Two Families of Effect Sizes

Two measures of effect sizes dominate the meta-analytic literature. When the primary studies in question compare two groups, either through experimental (treatment) versus control group comparisons or through orthogonal contrasts, the effect-size estimate often is expressed as some form of standardized difference between the group means. When two continuous variables are related, the Pearson product-moment correlation coefficient is most often used. We discuss each effect size family in turn.

5.2.1 The Standardized Mean Difference Family

Suppose that the data arise from a series of k independent studies, each of which compares a treatment or experimental group (E) with a control group (C). Let $Y_{E_{ij}}$ and $Y_{C_{ij}}$ denote the jth observations in the experimental and control groups of the ith study. Define the population standardized mean difference as

$$\delta_i = \frac{\mu_{E_i} - \mu_{C_i}}{\sigma_i}, \tag{5.1}$$

where μ_{E_i} and μ_{C_i} are the population means for the experimental and control groups in the ith study, respectively, and σ_i is the population standard deviation for the mean difference in the ith study.

Glass (1976) proposed that δ_i be estimated using

$$\hat{\Delta}_i = \frac{\bar{Y}_{E_i} - \bar{Y}_{C_i}}{S_{C_i}}, \tag{5.2}$$

where $\bar{Y}_{E_i}$ and $\bar{Y}_{C_i}$ are the respective sample means for the experimental and control groups in the ith study, and S_{C_i} is the control group sample standard deviation in the ith study. The estimated variance of $\hat{\Delta}_i$ is given by

$$\text{Var}\left(\hat{\Delta}_i\right) = \frac{n_{E_i} + n_{C_i}}{n_{E_i} n_{C_i}} + \frac{\hat{\Delta}_i^2}{2\left(n_{C_i} - 1\right)}, \tag{5.3}$$

where n_{E_i} and n_{C_i} are the respective sample sizes for the experimental and control groups in the ith study (Hedges, 1981, p. 112). Glass's estimator is appropriate if more than one experimental group is to be compared to a common control group (see Chapter 10) or if the experimental and control group population standard deviations almost surely will differ. For example, the experimental and control group population standard deviations would differ if the treatment affects the spread of scores as well as the magnitude of scores. However, Hedges and Olkin (1985) have pointed out that it is often reasonable to assume that the population variances do not differ even if the sample variances do differ. The equal variance assumption is especially likely to hold when only two groups are compared. Under the equal variance assumption, a more precise estimator of δ_i can be obtained by pooling the variances for the two groups.

If the population standard deviation $\sigma_i^E = \sigma_i^C = \sigma_i$ is known, you can estimate δ_i using

$$d_i = \frac{\overline{Y}_{E_i} - \overline{Y}_{C_i}}{\sigma_i} . \qquad (5.4)$$

However, the population standard deviation σ_i usually is not known and must be estimated. The maximum likelihood estimator (MLE) of σ_i is

$$S_{M_i} = \sqrt{\frac{\sum_{j=1}^{n_{E_i}}\left(Y_{E_{ij}} - \overline{Y}_{E_j}\right)^2 + \sum_{j=1}^{n_{C_i}}\left(Y_{C_{ij}} - \overline{Y}_{C_i}\right)^2}{n_{E_i} + n_{C_i}}} = \sqrt{\frac{\left(n_{E_i} - 1\right)S_{E_i}^2 + \left(n_{C_i} - 1\right)S_{C_i}^2}{n_{E_i} + n_{C_i}}} , \qquad (5.5)$$

where $S_{E_i}^2$ and $S_{C_i}^2$ are the respective sample variances for the experimental and the control groups in the ith study (Hedges and Olkin, 1985, page 82). Cohen (1969) proposed that δ_i be estimated using

$$d_i = \frac{\overline{Y}_{E_i} - \overline{Y}_{C_i}}{S_{M_i}} . \qquad (5.6)$$

When both the experimental and control group sample sizes are large, the estimated variance of d_i is given by

$$\text{Var}(d_i) = \left(\frac{n_{E_i} + n_{C_i}}{n_{E_i} n_{C_i}} + \frac{d_i^2}{2\left(n_{E_i} + n_{C_i} - 2\right)}\right) . \qquad (5.7)$$

Hedges (1981) used the pooled sample standard deviation

$$S_{POOLED_i} = \sqrt{\frac{\sum_{j=1}^{n_{E_i}}\left(Y_{E_{ij}} - \bar{Y}_{E_j}\right)^2 + \sum_{j=1}^{n_{C_i}}\left(Y_{C_{ij}} - \bar{Y}_{C_i}\right)^2}{n_{E_i} + n_{C_i} - 2}} = \sqrt{\frac{\left(n_{E_i} - 1\right)S_{E_i}^2 + \left(n_{C_i} - 1\right)S_{C_i}^2}{n_{E_i} + n_{C_i} - 2}} \qquad (5.8)$$

to estimate σ_i. Thus, Hedges' estimator for the standardized mean difference δ_i is

$$g_i = \frac{\bar{Y}_{E_i} - \bar{Y}_{C_i}}{S_{POOLED_i}}. \qquad (5.9)$$

When both the experimental and control group sample sizes are large, the estimated variance of g_i is given by

$$\text{Var}(g_i) = \frac{n_{E_i} + n_{C_i}}{n_{E_i} n_{C_i}} + \frac{g_i^2}{2\left(n_{E_i} + n_{C_i} - 2\right)} \qquad (5.10)$$

(Hedges 1981, p. 112).

All three estimators of δ_i are biased, and the bias can be a serious problem when both the experimental and control group sample sizes are small. Under the equal variance assumption, Cohen's d and Hedges' g are more precise estimators than Glass's $\hat{\Delta}$, and Hedges' g has smaller sample variance than Cohen's d. Hedges (1982) provided an exact correction factor for this sample bias

$$C(\upsilon_i) = \frac{\Gamma\left(\dfrac{\upsilon_i}{2}\right)}{\sqrt{\dfrac{\upsilon_i}{2}}\ \Gamma\left(\dfrac{\upsilon_i - 1}{2}\right)}, \qquad (5.11)$$

where $\upsilon_i = n_{E_i} + n_{C_i} - 2$ are the degrees of freedom and $\Gamma(.)$ is the gamma function (Hedges 1982, p. 492). Thus, an unbiased estimator of δ_i is given by

$$g_{U_i} = C(\upsilon_i)g_i. \qquad (5.12)$$

When $n_{E_i} = n_{C_i}$, g_{U_i} is not only an unbiased estimator of δ_i, it is also the unique minimum variance unbiased estimator of δ_i (see Hedges, 1981).

When both the experimental and control group sample sizes are large, the estimated variance of g_{U_i} is given by

$$\mathrm{Var}\left(g_{U_i}\right) = \frac{n_{E_i} + n_{C_i}}{n_{E_i} n_{C_i}} + \frac{g_{U_i}^2}{2\left(n_{E_i} + n_{C_i}\right)} \tag{5.13}$$

(Hedges and Olkin 1985, pp. 86). This chapter focuses on the unbiased estimator g_U proposed by Hedges (1981) because it has nice statistical properties. Appendix 5.1 shows how to convert the other estimators to g_U.

5.2.2 The Correlation Family

Suppose that the data arise from a series of k independent studies, each of which measures the linear relation between two continuous variables, say X and Y. The population correlation coefficient for the ith study is

$$\rho_i = \frac{\mathrm{Cov}\left(X_i,\ Y_i\right)}{\sigma_{X_i} \sigma_{Y_i}}, \tag{5.14}$$

where $\mathrm{Cov}\left(X_i,\ Y_i\right)$ is the population covariance of X and Y in the ith study, and σ_{X_i} and σ_{Y_i} are the respective population standard deviations for X and Y in the ith study.

The sample estimator of ρ_i is the Pearson product-moment correlation coefficient

$$r_i = \frac{\sum_{j=1}^{n_i}\left(X_{ij} - \bar{X}_i\right)\left(Y_{ij} - \bar{Y}_i\right)}{\sqrt{\left[\sum_{j=1}^{n_i}\left(X_{ij} - \bar{X}_i\right)^2\right]\left[\sum_{j=1}^{n_i}\left(Y_{ij} - \bar{Y}_i\right)^2\right]}}, \tag{5.15}$$

where $\bar{X}_i$ and $\bar{Y}_i$ are the respective mean values for the independent and dependent variables in the ith study.

5.2.3 Relationship between the Two Families of Effect Sizes

In a meta-analytical review, however, it is often the case that the effect size estimates will be a mixture of correlation coefficients and standardized mean differences. Although correlation coefficients cannot always be converted to standardized mean differences, in some special cases they can. When one of the variables (X) being correlated is dichotomous (that is, it can take on only two values), and the other variable (Y) is continuous, you can compute the point biserial correlation r_{pb} between the two variables. Some examples of dichotomous variables include male or female, smoker or nonsmoker, and the presence or absence of any characteristic. Although dichotomous variables are not quantitative, you can arbitrarily assign the two categories of the dichotomous variable the values 1 or 0 to reflect the presence or absence of the characteristic under consideration.[1] For example, you can assign the value 1 to all smokers (presence of smoking) and the value 0 to all nonsmokers (absence of smoking). The choice of which category is coded 1 and which is coded 0 will affect the direction but not the magnitude of the point-biserial correlation r_{pb}. The direction of r_{pb} will be positive if the average score on variable Y is greater for category 1 than for category 0, and it will be negative if the average score on variable Y is greater for category 0 than for category 1.

The standard deviation of the dichotomous variable X is determined by the proportion of the total sample size in each of the two categories:

$$S_X = \sqrt{pq},\qquad(5.16)$$

where p is the proportion of the total sample in category 1 and $q = 1 - p$ is the proportion in category 0 (Cohen & Cohen, 1983, p. 38). When X is dichotomous, the Pearson product-moment correlation coefficient in Equation 5.15 reduces to

$$r_{pb_i} = \frac{\left(\bar{Y}_{1_i} - \bar{Y}_{0_i}\right)\sqrt{p_i q_i}}{S_{Y_i}},\qquad(5.17)$$

---

[1] Actually, you can use any two values for the dichotomous X variable. For example, the value of the point-biserial correlation will be the same if you use the values 5 and 2 instead of the values 1 and 0.

where $\bar{Y}_{1_i}$ and $\bar{Y}_{0_i}$ are the Y sample means for categories 1 and 0 in the ith study,

respectively, and $S_{Y_i} = \sqrt{\sum_{i=1}^{n} \frac{(Y_i - \bar{Y})^2}{n}}$, where $\bar{Y}$ is the mean for variable Y

(combining categories 1 and 0) and n is the total sample size (Cohen & Cohen,

1983, p. 38). It is not difficult to see from Equation 5.17 that r_{pb} is closely related

to the standardized mean difference measures. Appendix 5.2 shows how to obtain

r_{pb} from other estimators.

Example 5.1 Calculating standardized mean difference estimates

This example shows how to use SAS software to compute the various

standardized mean difference estimates. Josephson (1987) conducted a study on

media-related aggression in children. Third-grade boys were first shown either a

violent police action drama or an exciting nonviolent program about a boys'

motocross bike-racing team. The boys then played a game of floor hockey. Two

observers coded all acts of aggression (for example, pushing someone down,

hitting someone with a hockey stick, elbowing, poking, pinching, hair pulling,

name calling). The average number of aggressive acts for the 44 boys who saw the

violent program was 1.81 with a standard deviation of 2.56. The average number of

aggressive acts for the 22 boys who saw the nonviolent program was 1.05 with a

standard deviation of 1.11.

In this study, we have all of the necessary information to compute Glass's $\hat{\Delta}$,

Cohen's d, Hedges' g, Hedges' g_U, and the point-biserial correlation r_{pb} using the

SAS macro COMPEFF(*indata,outdata*) in Appendix 5.3. *Indata* and *outdata* are

the names of the input and output data sets, respectively. *Indata* contains six

variables: the experimental and control group sample sizes (NE and NC,

respectively), sample means (ME and MC, respectively), and sample standard

deviations (SE and SC, respectively). The following SAS code was used to
obtained the results in Output 5.1:

```
options nocenter nodate pagesize=54 linesize=80 pageno=1;
libname ch5 "d:\metabook\ch5\dataset";
data ch5.ex51;
   input ne nc me mc se sc;
cards;
44 22 1.81 1.05 2.56 1.11
%compeff(ch5.ex51,ex51out);
data ex51out;
  set ex51out;
  if (type > 1);
proc format;
 value aa 0 = "Glass's Delta  "
          1 = "Cohen's d      "
          2 = "Hedges' g      "
          3 = "Hedges' gu     "
          4 = "rpb            ";
proc print data=ex51out noobs;
   var type estimate lower upper;
   title;
   format type aa. estimate 5.3 lower upper 6.4;
run;
```

Output 5.1 *Standardized mean difference estimates from Example 5.1*

TYPE	ESTIMATE	LOWER	UPPER
Hedges' g	0.347	-.1687	0.8619
Hedges' gu	0.343	-.1726	0.8576
rpb	0.164	-.0816	0.3903

As you can see in Output 5.1, Hedges' g is 0.347, Hedges' g_U is 0.343, and
r_{pb} is .164. The difference between g and g_U is small, indicating that the
correction factor for bias makes little difference. Cohen (1988) has offered
conventional values for "small," "medium," and "large" effects. For the
standardized mean difference, these conventional values are 0.2, 0.5, and 0.8,
respectively. For the correlation coefficient, these conventional values are .1, .3,
and .5, respectively. The estimates obtained for Josephson's (1987) study are
between Cohen's conventional values for small and medium effects — typical
values for social science studies (see Cohen, 1988). The estimates for this study do

not differ from zero, however, because all of the confidence intervals include the value zero.

5.3 Converting Test Statistics to Effect-Size Estimates and Converting Effect-Size Estimators from One Type to Another

In actual meta-analytical work, many studies do not report effect-size estimates. If these studies do report a two-group t-test statistic or an F-test statistic with 1 degree of freedom in the numerator, an effect-size estimate can be obtained.[2] It is not difficult to see that under the null hypothesis the population effect size is zero, the variances for Glass's $\hat{\Delta}$ (Equation 5.3), Cohen's d (Equation 5.7), Hedges' g (Equation 5.10), and Hedges' g_U (Equation 5.13) all reduce to

$$\frac{n_E + n_C}{n_E n_C}.$$
(5.18)

Appendix 5.4 shows how to obtain g, g_U, and r_{pb} from a t test. The SAS macro in Appendix 5.5 computes the effect-size estimates g, g_U, and r_{pb} from a t test.

If studies do report effect-size estimates, they are often of different types. Effect-size estimates need to be of the same type before they can be combined. The SAS macro in Appendix 5.5 converts effect-size estimates from one type to another.

---

[2] Some studies in a meta analytical review may only contain information about the direction and/or statistical significance of the hypothesis test. You can still use this information to get a lower limit for the effect-size estimate (Rosenthal, 1994). However, this approach is conservative. Bushman and Wang (1996) developed an alternative procedure for handling missing effect-sizes estimates that is described in Chapter 7.

Example 5.2 Converting test statistics to effect-size estimates

Malaria is the most important parasitic disease that affects humans because it is endemic in 103 countries, in which over half of the world's population live (Alonso et al., 1994). Each year, malaria claims the lives of between 1 and 3 million people, most of whom are young children and pregnant women in sub-Saharan Africa (Stuerchler, 1989). Alonso et al. (1994) conducted a randomized double-blind, placebo-controlled trial of the effects of SPf66 vaccine on *Plasmodium falciparum* malaria in children who live in southern Tanzania. In this part of the world, about 80% of infants are infected with malaria by age six months, and most people receive more than 300 infective mosquito bites per year (Smith et al., 1993).

In the study by Alonso et al. (1994), children aged one through five years were randomly assigned to receive three vaccine or placebo doses over a 26-week period. Four weeks after the third dose, anti-SPf66 antibodies were measured. The mean response was significantly higher for the 74 children in the vaccine group ($\bar{Y} = 8.3$) than for the 91 children in the placebo group ($\bar{Y} = 0.7$), $t(163) = 3.36$, $p < .001$. The research report did not, however, include the standard deviations for the two groups.

The SAS macro COVTEFST(*indata,outdata*) in Appendix 5.5 was used to obtain Hedges' g, Hedges' g_U, and point-biserial correlation r_{pb} for this example. *Indata* and *outdata* are the names of the input and output data sets, respectively. *Indata* contains four variables: ESTYPE, NE, NC, and STAT. ESTYPE is a categorical variable with values 1, 2, 3, 4, 5, and 6, depending on whether the study reports Cohen's d (ESTYPE=1), Hedges' g (ESTYPE=2), Hedges' g_U (ESTYPE=3), point-biserial correlation r_{pb} (ESTYPE=4), a t test statistic (ESTYPE=5), or an F test statistic with 1 degree of freedom in the numerator (ESTYPE=6). NE and NC are the sample sizes for the experimental and control groups, respectively. STAT is the numerical value of the effect-size estimate or test statistic that is specified by variable ESTYPE. For example, STAT is the value of the point-biserial correlation if ESTYPE=4, whereas STAT is the value of the t test statistic if ESTYPE=5. The following SAS code produced the results in Output 5.2:

```
options nodate nocenter pagesize=54 linesize=80 pageno=1;
data ex52;
    input estype ne nc stat;
cards;
5 74 91 3.36
%covtefst(ex52,ex52out);
proc format;
  value aa 2 = "Hedges' g "
           3 = "Hedges' gu"
           4 = "rpb       ";
data ex52out;
    set ex52out;
    if (2 <= type <= 4);
proc print data=ex52out noobs;
    var type estimate lower upper;
    title;
    format type aa. estimate 5.3 lower upper 6.4;
run;
```

Output 5.2 *Converting test statistics to effect-size estimates for Example 5.2*

TYPE	ESTIMATE	LOWER	UPPER
Hedges' g	0.526	0.2139	0.8380
Hedges' gu	0.524	0.2116	0.8355
rpb	0.255	0.1058	0.3920

As you can see from Output 5.2, Hedges' g is 0.526, Hedges' g_U is 0.524, and the point-biserial correlation r_{pb} is .255. The difference between g and g_U is small, indicating that the correction factor for bias makes little difference. All effect-size estimates are similar to Cohen's (1988) conventional values for medium-sized effects. Thus, the vaccine had a moderate effect on producing anti-SPf66 antibodies. All of the effect-size estimates differ from zero because none of the confidence intervals include the value zero.

5.4 Combining Sample Standardized Mean Differences

In a meta-analytic review, you want to combine the effect-size estimates from k studies. Suppose that each of the k studies has a large enough sample size for classical large sample methods that assume normality to be valid. Also suppose that the effect-size estimates are homogeneous (see Chapter 8, Section 8.4.1.1, for a description of homogeneity tests in fixed-effects models). Let $\hat{\delta}_i$ and $\text{Var}(\hat{\delta}_i)$ represent the effect-size measure and its corresponding large sample variance estimator in the ith study. For example, $\hat{\delta}_i$ could be $\hat{\Delta}_i$ (Equation 5.2), d_i (Equation 5.6), g_i (Equation 5.9), or g_{U_i} (Equation 5.12). We recommend the use of g_{U_i} because it is an unbiased estimator. In a fixed-effects meta-analysis, each effect-size estimate can be weighted by the inverse of its variance to obtain a pooled estimate of the combined estimator for the effect size (see Chapter 8 for a discussion of fixed-effects models). The pooled estimator of the population effect size δ across the k studies is

$$\hat{\delta}_+ = \frac{\sum_{i=1}^{k} w_i \hat{\delta}_i}{\sum_{i=1}^{k} w_i}, \tag{5.19}$$

where each effect-size estimate is weighted by the inverse of its variance

$$w_i = \frac{1}{\text{Var}(\hat{\delta}_i)}. \tag{5.20}$$

The + subscript is standard statistical notation that indicates that the effect-size estimates for the individual studies have been combined across all k studies. For example, to obtain a weighted average based on Hedges' unbiased estimator (that is, g_{U_+}), replace $\hat{\delta}_i$ with g_{U_i} (Equation 5.12) in Equation 5.19, and replace $\text{Var}(\hat{\delta}_i)$ with $\text{Var}(g_{U_i})$ (Equation 5.13) in Equation 5.20.

The variance of $\hat{\delta}_+$ is

$$\mathrm{Var}\left(\hat{\delta}_+\right) = \frac{\sum\limits_{i=1}^{k} w_i^2 \mathrm{Var}\left(\hat{\delta}_i\right)}{\left(\sum\limits_{i=1}^{k} w_i\right)^2} = \frac{\sum\limits_{i=1}^{k} w_i^2\left(\dfrac{1}{w_i}\right)}{\left(\sum\limits_{i=1}^{k} w_i\right)^2} = \frac{\sum\limits_{i=1}^{k} w_i}{\left(\sum\limits_{i=1}^{k} w_i\right)^2} = \frac{1}{\sum\limits_{i=1}^{k} w_i}. \tag{5.21}$$

The $100(1-\alpha)\%$ confidence interval for δ is given by

$$\hat{\delta}_+ - Z_{\alpha/2}\sqrt{\mathrm{Var}\left(\hat{\delta}_+\right)} \le \delta \le \hat{\delta}_+ + Z_{\alpha/2}\sqrt{\mathrm{Var}\left(\hat{\delta}_+\right)}, \tag{5.22}$$

where $Z_{\alpha/2}$ is the two-sided critical value from the standard normal distribution.

***Example 5.3** Combining standardized mean difference estimates*

Example 5.3 comes from a meta-analysis on media-related aggression by Wood, Wong, and Chachere (1991). All of the studies included in the meta-analysis examined media effects on aggressive behavior in naturally occurring social interactions. The study used in Example 5.1 was included in the meta-analysis by Wood and her colleagues. In the Wood et al. meta-analysis, five studies reported means and standard deviations, four studies reported Hedges' g, and two studies reported F statistics with one degree of freedom in the numerator.[3] The data are given in Table 5.1.

---

[3] We thank Wendy Wood for providing the means, standard deviations, effect-size estimates, and test statistics for the studies that are included in the meta-analysis.

Table 5.1 *Wood et al. (1991) meta-analysis on the effects of violent media on aggression in naturally occurring social interactions*

Study	Violent media group			Control group			g	Test statistic
	Y_E	S_E	n_E	Y_C	S_C	n_C		
1	1.96	0.548	30	1.78	0.548	30		
2			14			14	-0.34	
3	13.05	11.38	20	4.00	11.38	10		
4	1.81	2.56	44	1.05	1.11	22		
5	4.19	2.87	16	2.59	2.87	16		
6	0.30	0.158	32	0.25	0.158	32		
7			48			48	-0.17	
8			26			26	-0.07	
9			17			34		$F(1, 32) = 7.18$
10			18.5			18.5		$F(1, 33) = 4.13$
11			5			5	1.04	

Note: g = Hedges' estimate of the population standardized mean difference.

To obtain an average estimate and confidence interval for the studies in the Wood et al. meta-analysis, you need the SAS macro COVTEFST(*indata,outdata*) in Appendix 5.5, the SAS macro WAVGMETA(*indata,outdata,kind,level*) in Appendix 5.6, and the following SAS code. In the SAS macro WAVGMETA, *indata* is the output data file name from the SAS macro COVTEFST. *Outdata* is the output data file name for the SAS macro WAVGMETA. *Kind* is a categorical variable with values 1, 2, 3, and 4, depending on whether the studies in the meta-analysis report Cohen's *d* (*kind*=1), Hedges' *g* (*kind*=2), Hedges' g_U (*kind*=3), or the point-biseral correlation r_{pb} (*kind*=4). All effect-size estimates reported should be converted to the same type, using the SAS macro in Appendix 5.3, before a weighted average of the estimates is obtained. It doesn't matter which type of estimator you use, however, because the SAS macro WAVGMETA prints out results for three types of estimators: Hedges' g, Hedges' g_U, and point-biseral correlation r_{pb}. In Example 6.3, $kind=2$ because all studies reported Hedges' g. *Level* is the confidence level for the confidence interval based on a weighted average of the effect-size estimates (for example, use the value .95 to obtain a 95% confidence interval).

```
options nodate nocenter linesize=80 pagesize=54 pageno=1;
libname ch5 "d:\metabook\ch5\dataset";
%covtefst(ch5.ex53,temp);
%wavgmeta(temp,combine1,2,0.95);
%covtefst(ch5.ex53,temp);
%wavgmeta(temp,combine2,3,0.95);
%covtefst(ch5.ex53,temp);
%wavgmeta(temp,combine3,4,0.95);
proc format;
    value aa 2 = "Hedges' g+ "
             3 = "Hedges' gu+"
             4 = "rpb+       ";
data ex53out;
    set combine1 combine2 combine3;
    keep type estimate lower upper;
proc print data=ex53out noobs;
    title;
    format type aa. estimate 5.3 lower upper 5.3;
run;
```

Output 5.3 includes results that are based on the following estimators: Hedges' g, Hedges' g_U, and the point-biserial correlation r_{pb}. As you can see in Output 5.3, none of the confidence intervals includes the value zero. The difference between g_+ and g_{U_+} is small, indicating that the correction factor for bias makes little difference. All estimates are between Cohen's (1988) conventional values for small- and medium-sized effects. Thus, media violence increased aggression in viewers.

Output 5.3 *Combining standardized mean difference estimates for meta-analysis on media-related aggression*

TYPE	ESTIMATE	LOWER	UPPER
Hedges' g+	0.262	0.086	0.437
Hedges' gu+	0.257	0.081	0.432
rpb+	0.186	0.098	0.268

5.5 Combining Sample Correlation Coefficients

One estimator of the population correlation coefficient ρ is the average of the sample correlation coefficients. Combining sample correlation coefficients, however, is complicated by the fact that the distribution of the correlation coefficient is not normal when $\rho \neq 0$. To remedy this problem, Fisher (1921) proposed a transformation from r to a quantity z that is approximately normally distributed with approximate variance $1/(n-3)$. The relation between r and z is given by

$$z = \frac{1}{2}\ln\frac{1+r}{1-r},$$

(5.23)

where ln(.) is the natural logarithmic function.

Suppose that the sample correlations are homogeneous (see Chapter 8, Section 8.4.1.1 for a discussion that tests this assumption). To obtain a weighted average of the sample correlations, r_+, first obtain a weighted average of the corresponding z values

$$z_+ = \frac{\sum_{i=1}^{k}(n_i-3)z_i}{\sum_{i=1}^{k}(n_i-3)},$$

(5.24)

where n_i is the sample size in the ith study. The estimated variance for z_+ is

$$
\begin{aligned}
\mathrm{Var}(z_+) &= \mathrm{Var}\left(\frac{\sum_{i=1}^{k}(n_i-3)z_i}{\sum_{i=1}^{k}(n_i-3)} \right) = \frac{\sum_{i=1}^{k}(n_i-3)^2 \times \mathrm{Var}(z_i)}{\left(\sum_{i=1}^{k}(n_i-3)\right)^2} \\[2em]
&= \frac{\sum_{i=1}^{k}(n_i-3)^2 \times \left(\frac{1}{n_i-3}\right)}{\left(\sum_{i=1}^{k}(n_i-3)\right)^2} = \frac{\sum_{i=1}^{k}(n_i-3)}{\left(\sum_{i=1}^{k}(n_i-3)\right)^2} \\[2em]
&= \frac{1}{\sum_{i=1}^{k}(n_i-3)} = \frac{1}{N-3k} ,
\end{aligned}
\tag{5.25}
$$

where $N = n_1 + \ldots + n_k$. The transformation in Equation 5.26 is used to obtain

$$
r_+ = \frac{\exp\{2z_+\}-1}{\exp\{2z_+\}+1} .
\tag{5.26}
$$

Two reasons exist for the choice of (n_i-3) as the weight in Equation 5.23. First, (n_i-3) is the inverse of the variance for Fisher's z transformation; effect-size estimates are often weighted by the inverse of their variance. Second, z_+ is an unbiased estimator of

$$
\zeta = \frac{1}{2}\ln\frac{1+\rho}{1-\rho} .
\tag{5.27}
$$

To obtain a $100(1-\alpha)\%$ confidence interval for the population correlation coefficient ρ, you first obtain a $100(1-\alpha)\%$ confidence interval for the population z transformation parameter ζ,

$$
z_+ - Z_{\alpha/2}\frac{1}{\sqrt{N-3k}} \le \zeta \le z_+ + Z_{\alpha/2}\frac{1}{\sqrt{N-3k}} ,
\tag{5.28}
$$

where $N = n_1 + \ldots + n_k$. You can then use the transformation in Equation 5.26 to obtain the upper and lower confidence interval bounds for ρ from the upper and lower confidence interval bounds for ζ.

Example 5.4 *Combining Pearson product-moment correlation coefficients*

The bone mass that is developed earlier in life is an important determinant of osteoporotic fractures later in life. Calcium intake is one factor that is suggested to be related to bone mass in young and middle-aged adults (Cummings, Kelsey, Nevitt, & O'Dowd, 1985). Welten, Kemper, Post, and Van Staveren (1995) conducted a meta-analysis of studies that examined the relation between calcium intake and bone mass in people between the ages of 18 and 50. The results from 20 cross-sectional studies that reported Pearson product-moment correlation coefficients are reported in Table 5.2.

Table 5.2 *Welten et al. (1995) meta-analysis of the relation between calcium intake and bone mass in young and middle aged adults*

Study	n	r
1	86	.29
2	88	.00
3	60	.24
4	183	.21
5	37	.15
6	101	.10
7	300	.07
8	248	.00
9	30	-.16
10	89	.00
11	182	.08
12	161	.21
13	296	.33
14	55	.26
15	114	.19
16	173	.03
17	249	.11
18	98	-.03
19	189	.00
20	84	.03

To obtain an estimate and a 95% confidence interval for the population correlation coefficient for Example 5.4, you can use the SAS macros COVTEFST and WAVGMETA in Appendices 5.5 and 5.6, respectively, and the following SAS code. For computational convenience, both NE and NC are set equal to half of the total sample size before calling the SAS macro:

```
options nodate nocenter pagesize=54 linesize=80 pageno=1;
libname ch5 "d:\metabook\ch5\DATASET";
%covtefst(ch5.ex54,temp);
%wavgmeta(temp,ex54out,4,0.95);
proc format;
    value aa 7 = "Fisher's z+    "
             4 = "Correlation r+";
data ex54out;
    set ex54out;
    output;
    type = 7;
    estimate = 0.5*log((1+estimate)/(1-estimate));
    lower = 0.5*log((1+lower)/(1-lower));
    upper = 0.5*log((1+upper)/(1-upper));
    output;
proc print data=ex54out noobs;
    var type estimate lower upper;
    format type aa. estimate lower upper 5.3;
    title;
run;
```

Output 5.4 *Combining Pearson product-moment correlation coefficients for the meta-analysis on the relation between calcium intake and bone mass*

TYPE	ESTIMATE	LOWER	UPPER
Correlation r+	0.115	0.077	0.151
Fisher's z+	0.115	0.078	0.152

As you can see in Output 5.4, the estimate of the population correlation is $r_+ = .115$ with a 95% confidence interval ranging from .077 to .151. The confidence interval does not include the value zero. Thus, there is a small but significant positive correlation between calcium intake and bone mass in these studies.

5.6 Conclusions

This chapter focused on combining effect-size estimates based on continuous dependent variables. There are two families of effect sizes: the standardized mean difference family and the correlation family. When the independent variable is dichotomous, the effect-size estimate often is expressed as some form of standardized difference between the group means. Several estimators exist for the standardized mean difference, including Glass's $\hat{\Delta}$, Cohen's d, Hedges' g, Hedges' g_U, and the point-biserial correlation r_{pb}. Because it has nice statistical properties, we recommend the use of Hedges' g_U. When the independent variable is continuous, the Pearson product-moment correlation coefficient is most often used.

The SAS macro in Appendix 5.5 produces effect-size estimates that are based on statistical tests and/or other type of effect-size estimates. If some studies report only the direction and/or statistical significance of results, then you should use the methods presented in Chapter 7 instead.

Population effect sizes are often estimated by averaging the effect-size estimates from the individual studies, after each estimate has been weighted by the inverse of its variance. Weighting studies by the inverse of their variances gives studies with larger sample sizes more weight in the average estimate. The SAS macro in Appendix 5.6 produces an estimate and 95% confidence intervals for the standardized mean difference and the correlation coefficient based on this weighted average procedure.

In this chapter, we assumed that the effect-size estimates to be combined are homogeneous. Of course, the meta-analyst should test this assumption before combining effect-size estimates. Homogeneity tests are described in Chapters 8 and 9.

5.7 References

Alonso, P. L., Smith, T., Armstrong Schellenberg, J. R. M., et al. (1994). Randomised trial of efficacy of Spf66 vaccine against *Plasmodium faliparum* malaria in children in southern Tanzania. *Lancet, 344*, 1175–1181.

Bushman, B. J. & Wang, M. C. (1996). A procedure for combining sample standardized mean differences and vote counts to estimate population standardized mean difference in fixed effect models. *Psychological Methods, 1*, 66–80.

Cohen, J. (1969). *Statistical power analysis for the behavioral sciences*. New York: Academic Press.

Cohen, J. (1988). *Statistical power analysis for the behavioral sciences* (2nd ed.). Hillsdale, NJ: Erlbaum.

Cohen, J. & Cohen, P. (1983). *Applied multiple regression/correlation analysis for the behavioral sciences* (2nd ed.). Hillsdale, N. J.: Erlbaum.

Cummings, S. R., Kelsey, J. L., Nevitt, M. C., & O'Dowd, K. J. (1985). Epidemiology of osteoporosis and osteoporotic fractures. *Epidemiology Review, 7*, 178–208.

Fisher, R. A. (1921). On the 'probable error' *of* a coefficient of correlation deduced from a small sample. *Metron, 1*, 1–32.

Glass, G. V. (1976). Primary, secondary, and meta-analysis of research. *Educational Researcher, 5*, 3–8.

Hedges, L. V. (1981). Distribution theory for Glass's estimator of effect size and related estimators. *Journal of Educational Statistics, 6*, 107–128.

Hedges, L. V. (1982). Estimation of effect size from a series of independent experiments. *Psychological Bulletin, 92*, 490–499.

Hedges, L. V. & Olkin, I. (1985). Statistical methods for meta-analysis. New York: Academic Press.

Josephson, W. L. (1987). Television violence and children's aggression: Testing the priming, social script, and disinhibition predictions. *Journal of Personality and Social Psychology, 53*, 882–890.

Rosenthal, R. (1994). Parametric measures of effect size. In H. Cooper & L. V. Hedges (Eds.). The handbook of research synthesis (pp. 231–244). New York: Russell Sage Foundation.

Smith, T., Charlwood, J. D., Kihonda, J., et al. (1993). Absence of seasonal variation in malaria parasitaemia in an area of intense seasonal transmission. *Acta Tropica, 54*, 55–72.

Stuerchler, D. (1989). Malaria, how much is there worldwide? *Parasitology Today, 5*, 39–40.

Welten, D. C., Kemper, H. C. G., Post, G. B., & Van Staveren, W. A. (1995). A meta-analysis of calcium intake in young and middle-aged females and males. *Journal of Nutrition, 125*, 2802–2813.

Wood, W., Wong, F. Y., & Chachere, J. G. (1991). Effects of media violence on viewers' aggression in unconstrained social interaction. *Psychological Bulletin, 109*, 371–383.

5.8 Appendices

Appendix 5.1: Formulas for Converting Cohen's d, Hedges' g, and the Point-Biseral Correlation to Hedges' g_u

The formula for obtaining r_{pb} from d is

$$r_{pb} = \sqrt{\frac{n_E n_C d^2}{n_E n_C d^2 + (n_E + n_C)^2}} \tag{5.29}$$

(see Cohen, 1988, page 24).

The formula for obtaining r_{pb} from g is

$$
\begin{aligned}
r_{pb} &= \sqrt{\frac{n_E n_C d^2}{n_E n_C d^2 + (n_E + n_C)^2}} \\
&= \sqrt{\frac{n_E n_C \left(\dfrac{n_E + n_C - 2}{n_E + n_C} g^2\right)}{n_E n_C \left(\dfrac{n_E + n_C - 2}{n_E + n_C} g^2\right) + (n_E + n_C)^2}} \\
&= \sqrt{\frac{n_E n_C g^2}{n_E n_C g^2 + (n_E + n_C)(n_E + n_C - 2)}}
\end{aligned}
\tag{5.30}
$$

because $d = \sqrt{\dfrac{n_E + n_C - 2}{n_E + n_C}} \times g$ (see Hedges & Olkin, 1985, page 82).

The formula for obtaining r_{pb} from g_U is

$$
\begin{aligned}
r_{pb} &= \sqrt{\frac{n_E n_C g^2}{n_E n_C g^2 + (n_E + n_C)(n_E + n_C - 2)}} \\
&= \sqrt{\frac{n_E n_C [g_U / C(n_E + n_C - 2)]^2}{n_E n_C [g_U / C(n_E + n_C - 2)]^2 + (n_E + n_C)(n_E + n_C - 2)}}
\end{aligned}
\tag{5.31}
$$

because $g = g_u / C(n_E + n_C - 2)$, as given in Equation 5.12, where $C(.)$ is given in Equation 5.11.

Appendix 5.2: Formulas for Converting Cohen's d, Hedges' g, and Hedges' g_u to Point-Biseral Correlations

The formula for obtaining g_U from g is

$$g_u = g \times C(n_E + n_C - 2),$$ (5.32)

where $C(.)$ is given in Equation 5.11 (see Hedges, 1982, p. 492).

The formula for obtaining g_U from d is

$$\begin{aligned} g_u &= g \times C(n_E + n_C - 2) \\ &= d \times \sqrt{\frac{n_E + n_C}{n_E + n_C - 2}} \times C(n_E + n_C - 2) \end{aligned}$$ (5.33)

because $g = d \times \sqrt{\dfrac{n_E + n_C}{n_E + n_C - 2}}$ (see Hedges & Olkin, 1985, page 82).

It is not difficult to verify that the formula for obtaining g_U from r_{pb} is

$$g_U = \left(\sqrt{\frac{(n_E + n_C)(n_E + n_C - 2)}{n_E n_C}} \right) \frac{r_{pb}}{\sqrt{1 - (r_{pb})^2}} C(n_E + n_C - 2),$$ (5.34)

where $C(.)$ is given in Equation 5.11.

Appendix 5.3: SAS Macro for Computing Effect-Size Estimates

```
%macro compeff(indata,outdata);
/*****************************************************/
/*   INDATA:   Input data set name                  */
/*   OUTDATA:  Output data set name                 */
/*                                                  */
/*   NE:   Experimental group sample size           */
/*   NC:   Control group sample size                */
/*   ME:   Experimental group sample mean           */
/*   MC:   Control group sample mean                */
/*   SE:   Experimental group standard deviation    */
/*   SC:   Control group standard deviation         */
/*                                                  */
/* OUTPUT DATA:                                     */
/*   TYPE:  0 = Glass' delta                        */
/*          1 = Cohen's d                           */
/*          2 = Hedges' g                           */
/*          3 = Hedges' gu                          */
/*          4 = Point biserial correlation rpb      */
/*   ESTIMATE:  Effect-size estimate                */
/*   STD:  Estimated standard deviation for         */
/*         effect-size estimate                     */
/*   LOWER:  Lower bound of 95% confidence interval */
/*   UPPER:  Upper bound of 95% confidence interval */
/*****************************************************/
data &outdata;
   set &indata;
   dfd = ne+nc;
   dfg = dfd-2;
/*****************************************************/
/*   Compute Glass's delta                          */
/*****************************************************/
   del = (me-mc)/sc;
   vdel = dfd/(ne*nc)+(del*del)/(2*nc-2);
   sdel = sqrt(vdel);
/*****************************************************/
/*   Compute Cohen's d                              */
/*****************************************************/
   sm = sqrt(((ne-1)*(se*se)+(nc-1)*(sc*sc))/dfd);
   dd = (me-mc)/sm;
   vdd = dfd/(ne*nc)+0.5*dd*dd/dfg;
   sdd = sqrt(vdd);
/*****************************************************/
/*   Compute Hedges' g                              */
/*****************************************************/
   sp = sqrt(((ne-1)*(se*se)+(nc-1)*(sc*sc))/dfg);
   gg = (me-mc)/sp;
   vgg = dfd/(ne*nc)+0.5*gg*gg/dfg;
   sgg = sqrt(vgg);
```

```
/*****************************************************/
/*  Compute Hedges' gu                            */
/*****************************************************/
   corrf=exp(lgamma(0.5*dfg)-lgamma(0.5*(dfg-1))
      -log(sqrt(0.5*dfg)));
   gu = gg * corrf;
   vgu = dfd/(ne*nc)+0.5*gu*gu/dfd;
   sgu = sqrt(vgu);
/*****************************************************/
/*  Compute point-biserial correlation rpb        */
/*****************************************************/
   rpb = sqrt((ne*nc*dd*dd)/(ne*nc*dd*dd+dfd*dfd));
   zrpb = 0.5*log((1+rpb)/(1-rpb));
   vzrpb = 1 / (dfd-3);
   szrpb = sqrt(vzrpb);
   type = 0;
   estimate = del;
   std = sdel;
   lower = del + probit(0.025) * sdel;
   upper = del + probit(0.975) * sdel;
   output;
   type = 1;
   estimate = dd;
   std = sdd;
   lower = dd + probit(0.025) * sdd;
   upper = dd + probit(0.975) * sdd;
   output;
   type = 2;
   estimate = gg;
   std = sgg;
   lower = gg + probit(0.025) * sgg;
   upper = gg + probit(0.975) * sgg;
   output;
   type = 3;
   estimate = gu;
   std = sgu;
   lower = gu + probit(0.025) * sgu;
   upper = gu + probit(0.975) * sgu;
   output;
   type = 4;
   estimate = rpb;
   std = szrpb;
   lower = zrpb + probit(0.025) * szrpb;
   upper = zrpb + probit(0.975) * szrpb;
   lower = (exp(2*lower)-1)/(exp(2*lower)+1);
   upper = (exp(2*upper)-1)/(exp(2*upper)+1);
   output;
   keep ne nc me mc se sc type estimate lower upper;
%mend compeff;
```

Appendix 5.4: Formulas for Obtaining Hedges' g, Hedges' g_U and the Point-Biseral Correlation from a t Test Statistic

The formula for obtaining Hedges' g from a t test statistic is

$$g = \frac{t\sqrt{n_E + n_C}}{\sqrt{n_E n_C}} \qquad (5.35)$$

because $\sqrt{\dfrac{n_E n_C}{n_E + n_C}}$ is the standard deviation of g under the null hypothesis that the population effect size is zero.

The formula for obtaining Hedges' g_U from a t test statistic is

$$g_U = g \times C(n_E + n_C - 2)$$
$$= \frac{t\sqrt{n_E + n_C}}{\sqrt{n_E n_C}} \times C(n_E + n_C - 2), \qquad (5.36)$$

where $C(.)$ is given in Equation 5.11.

The formula for obtaining a point-biseral correlation from a t test statistic is

$$r_{pb} = \sqrt{\frac{n_E n_C g^2}{n_E n_C g^2 + (n_E + n_C)(n_E + n_C - 2)}}$$
$$= \sqrt{\frac{n_E n_C \left(\frac{t\sqrt{n_E + n_C}}{\sqrt{n_E n_C}}\right)^2}{n_E n_C \left(\frac{t\sqrt{n_E + n_C}}{\sqrt{n_E n_C}}\right)^2 + (n_E + n_C)(n_E + n_C - 2)}} \qquad (5.37)$$
$$= \sqrt{\frac{t^2}{t^2 + (n_E + n_C - 2)}}.$$

Given the identity that $t^2 = F$ when there is 1 degree of freedom in the numerator of the F test, identical estimates can be obtained by substituting $\sqrt{F}$ for t in Equations 5.35 through 5.37.

Appendix 5.5: SAS Macro for Converting Test Statistics to Effect-Size Estimates and for Converting Effect-Size Estimators from One Type to Another

```
%macro covtefst(indata,outdata);
/***********************************************************/
/*   INDATA: Input data set name                          */
/*   OUTDATA: Output data set name                        */
/***********************************************************/
data &outdata;
   set &indata;
/***********************************************************/
/*   INPUT DATA:                                          */
/*     ESTYPE: Type of effect-size estimate or statistic  */
/*            ESTYPE = 1 for Cohen's d                    */
/*            ESTYPE = 2 for Hedges' g                    */
/*            ESTYPE = 3 for Hedges' gu                   */
/*            ESTYPE = 4 for Point-biserial correlation rpb */
/*            ESTYPE = 5 for t statistic                  */
/*            ESTYPE = 6 for F statistic                  */
/*                                                        */
/*     NE:  Experimental group sample size               */
/*     NC:  Control group sample size                    */
/*     STAT: STAT is effect-size estimate if STYPE = 0   */
/*           STAT is t statistic if STYPE = 1            */
/*           STAT is F statistic if STYPE = 2            */
/*                                                        */
/*   OUTPUT DATA:                                         */
/*     NE:  Experimental group sample size               */
/*     NC:  Control group sample size                    */
/*     TYPE: 1 = Cohen's d                                */
/*             2 = GG: Hedges' g                          */
/*             3 = GU: Hedges' gu                         */
/*             4 = Point-biserial correlation rpb         */
/*             5 = t statistic                            */
/*             6 = F statistic                            */
/*     ESTIMATE: Effect-size estimate                     */
/*     LOWER:   Lower bound for 95% confidence interval   */
/*     UPPER:   Upper bound for 95% confidence interval   */
/***********************************************************/
   dfd=nc+ne;
   dfg=nc+ne-2;
   corrf=exp(lgamma(0.5*dfg)-lgamma(0.5*(dfg-1))
      -log(sqrt(0.5*dfg)));
   select (estype);
   when (1) do;
```

```
/***********************************************************/
/*   Converting Cohen's d to Hedges' g, Hedges' gu,      */
/*   rpb, t statistic, and F statistic                   */
/***********************************************************/
      dd=stat;
      gg=sqrt(dfg/dfd)*dd;
      gu=corrf*gg;
      rpb=sqrt((nc*ne*dd*dd)/(nc*ne*dd*dd+dfd*dfd));
      tstat = gg * sqrt(ne * nc / dfd);
      fstat = tstat * tstat;
   end;
   when (2) do;
/***********************************************************/
/*   Converting Hedges' g to Cohen's d, Hedges' gu,      */
/*   rpb, t statistic, and F statistic                   */
/***********************************************************/
      gg=stat;
      dd=gg*sqrt(dfd/dfg);
      gu=corrf*gg;
      rpb=sqrt((ne*nc*dd*dd)/(ne*nc*dd*dd+dfd*dfd));
      tstat = gg * sqrt(ne * nc / dfd);
      fstat = tstat * tstat;
   end;
   when (3) do;
/***********************************************************/
/*   Converting Hedges' gu to Cohen's d, Hedges' g,      */
/*   rpb,  t statistic, and F statistic                  */
/***********************************************************/
      gu=stat;
      gg=gu/corrf;
      dd=gg*sqrt(dfd/dfg);
      rpb=sqrt((ne*nc*dd*dd)/(ne*nc*dd*dd+dfd*dfd));
      tstat = gg * sqrt(ne * nc / dfd);
      fstat = tstat * tstat;
   end;
   when (4) do;
/***********************************************************/
/*   Converting rpb to Cohen's d, Hedges' g, Hedges' gu, */
/*    t statistic, and F statistic                       */
/***********************************************************/
      rpb=stat;
      dd=(dfd*rpb)/sqrt(nc*ne*(1-rpb*rpb));
      gg=dd*sqrt(dfg/dfd);
      gu=gg*corrf;
      tstat = gg * sqrt(ne * nc / dfd);
      fstat = tstat * tstat;
   end;
```

```
/***********************************************************/
/*  Converting t Statistics to Cohen's d, Hedges' g,      */
/*  Hedges' gu, and F statistics                          */
/***********************************************************/
   when (5) do;
      tstat = stat;
      fstat = tstat * tstat;
      dd = tstat * dfd / sqrt(ne*nc*dfg);
      gg=sqrt(dfg/dfd)*dd;
      gu=corrf*gg;
      rpb=sqrt((nc*ne*dd*dd)/(nc*ne*dd*dd+dfd*dfd));
   end;
/***********************************************************/
/*  Converting F statistics to Cohen's d, Hedges' g,      */
/*  Hedges' gu, and t statistics                          */
/***********************************************************/
   when (6) do;
      fstat = stat;
      tstat = sqrt(fstat);
      dd = tstat * dfd / sqrt(ne*nc*dfg);
      gg=sqrt(dfg/dfd)*dd;
      gu=corrf*gg;
      rpb=sqrt((nc*ne*dd*dd)/(nc*ne*dd*dd+dfd*dfd));
   end;
   otherwise put "estype can only be 1, 2, 3, 4, 5, 6";
   end;
   vdd = dfd/(nc*ne) + (dd * dd) / (2 * dfg);
   vgg = dfd/(nc*ne) + (gg * gg) / (2 * dfg);
   vgu = dfd/(nc*ne) + (gu * gu) / (2 * dfd);
   zrpb = 0.5 * log((1+rpb)/(1-rpb));
   vzrpb = 1 / (dfd - 3);
   type = 1;
   estimate = dd;
   std = sqrt(vdd);
   lower = dd + probit(0.025) * std;
   upper = dd + probit(0.975) * std;
   output;
   type = 2;
   estimate = gg;
   std = sqrt(vgg);
   lower = gg + probit(0.025) * std;
   upper = gg + probit(0.975) * std;
   output;
   type = 3;
   estimate = gu;
   std = sqrt(vgu);
   lower = gu + probit(0.025) * std;
   upper = gu + probit(0.975) * std;
   output;
   type = 4;
   estimate = rpb;
   std = sqrt(vzrpb);
   lower = zrpb + probit(0.025) * std;
```

```
      upper = zrpb + probit(0.975) * std;
      lower = (exp(2*lower)-1)/(exp(2*lower)+1);
      upper = (exp(2*upper)-1)/(exp(2*upper)+1);
      output;
      type = 5;
      estimate = tstat;
      lower = .;
      upper = .;
      output;
      type = 6;
      estimate = fstat;
      std = . ;
      lower = .;
      upper = .;
      keep type estimate std lower upper ne nc;
  run;
  %mend covtefst;
```

Appendix 5.6: SAS Macro for Computing a Weighted Average of Effect-Size Estimates

```
%macro wavgmeta(indata,outdata,kind,level);
/*************************************************************/
/*    INDATA: Input data set name                          */
/*    OUTDATA: Output data set name                        */
/*    KIND:  Type of effect-size to be combined            */
/*    LEVEL: Confidence level for confidence interval      */
/*                                                         */
/*    Input Data File:                                     */
/*       The input data file for this macro is the output  */
/*       data file from SAS Macro COVTEFST                 */
/*       TYPE: Type of effect-size estimate to be combined */
/*          TYPE = 1  (Cohen's d)                          */
/*          TYPE = 2  (Hedges' g)                          */
/*          TYPE = 3  (Hedges' gu)                         */
/*          TYPE = 4  (Correlation coefficient)            */
/*       ESTIMATE:  Effect-size estimate                   */
/*       STD:  Standard deviation of effect-size estimate  */
/*       NE:  Experimental group sample size               */
/*       NC:  Control group sample size                    */
/*       LOWER:  Lower bound of 95% confidence interval    */
/*       UPPER:  Upper bound of 95% confidence interval    */
/*                                                         */
/*    Output Data File:                                    */
/*                                                         */
/*       Variables:                                        */
/*       TYPE:  Type of effect-size estimate               */
/*       WGTAVG:  Weighted average of effect-size estimates*/
/*       STDWAVG: Estimated standard deviation of average  */
/*                effect-size estimate                     */
/*       LOWER:  Lower bound of 95% confidence interval    */
/*       UPPER:  Upper bound of 95% confidence interval    */
/*************************************************************/
data &indata;
   set &indata;
   if (type = &kind);
   alpha = 1-&level;
   dfd = ne + nc;
   dfg = ne + nc - 2;
   select (type);
      when (1) do;
         vv = dfd/(ne*nc)+(estimate*estimate)/(2*dfg);
      end;
      when (2) do;
         vv = dfd/(ne*nc)+(estimate*estimate)/(2*dfg);
      end;
      when (3) do;
         vv = dfd/(ne*nc)+(estimate*estimate)/(2*dfd);
      end;
```

```
      when (4) do;
          vv = 1 / (dfd - 3);
      end;
   end;
   effect = estimate;
   ww = 1 / vv;
   wweff = ww * effect;
   keep type effect vv ww wweff alpha;
proc means data=&indata sum mean noprint;
   var type ww wweff alpha;
   output out=&outdata sum = stype sww swweff
      mean = mtype mww mwweff malpha;
data &outdata;
   set &outdata;
   type = mtype;
   estimate = swweff / sww;
   stdwavg =sqrt(1/sww);
   lower = estimate + probit(0.5*malpha) * stdwavg;
   upper = estimate + probit(1-0.5*malpha) * stdwavg;
   if (type = 4) then do;
       lower = (exp(2*lower)-1)/(exp(2*lower)+1);
       upper = (exp(2*upper)-1)/(exp(2*upper)+1);
   end;
   keep type estimate lower upper;
run;
%mend wavgmeta;
```

chapter 6

Vote-Counting Procedures in Meta-Analysis

6.1 Introduction

The meta-analyst has access to at least one of four types of information from a primary research report: (a) information that can be used to compute an effect-size estimate (for example, means, standard deviations, test statistic values), (b) information about whether the hypothesis test found a statistically significant relation between the independent and dependent variables and the direction of the relation (for example, a significant positive mean difference), (c) information about only the direction of the relation between the independent and dependent variables (for example, a positive mean difference), and (d) no information about the relation between the independent and dependent variables. These types are rank ordered, from most to least, in terms of the amount of information that they contain (Hedges, 1986). You should use effect-size procedures for the studies that contain enough information to calculate an effect-size estimate (see Chapter 5). Use Vote-counting procedures for the studies that do not contain enough information to calculate an effect-size estimate, but do contain information about the direction and/or statistical significance of results. This chapter focuses on vote-counting procedures (see Bushman, 1994, for a more detailed discussion of vote-counting procedures).

We recommend that you never use vote-counting procedures alone (unless none of the studies in a meta-analysis reports effect-size estimates). We recommend that you use vote-counting procedures in conjunction with effect-size procedures. This chapter is primarily a "stepping stone" for the next chapter. Chapter 7 describes how sample effect sizes and vote-counts can be combined to estimate population effect sizes.

6.2 The Conventional Vote-Counting Procedure

Light and Smith (1971) were the first to propose a formal procedure for "taking a vote" of study results:

> *All studies which have data on a dependent variable and a specific independent variable of interest are examined. Three possible outcomes are defined. The relationship between the independent variable and the dependent variable is either significantly positive, significantly negative, or there is no specific relationship in either direction. The number of studies falling into each of these three categories is then simply tallied. If a plurality of studies falls into any one of these three categories, with fewer falling into the other two, the modal category is declared the winner. This modal category is then assumed to give the best estimate of the direction of the true relationship between the independent and dependent variable. (p. 433).*

The conventional vote-counting procedure proposed by Light and Smith (1971) has been criticized on several grounds. First, as Light and Smith noted, the conventional vote-counting procedure does not incorporate sample size into the vote. It is well known that as sample size increases, the probability of finding a statistically significant relation between the independent and dependent variables also increases. Second, although the conventional vote-counting procedure allows you to determine which modal category is the "winner", it does not allow you to determine what the margin of victory is (that is, it does not provide an effect-size estimate). Third, Hedges and Olkin (1980) have shown that the conventional vote-counting procedure has very low power for the range of sample sizes and effect sizes most common in the social sciences. The vote-counting procedures proposed by Hedges and Olkin (1980) overcome the problems that are associated with the conventional vote-counting procedure proposed by Light and Smith. The vote-counting procedures proposed by Hedges and Olkin are described in this chapter, and SAS code is given to implement these procedures. We describe vote-counting procedures based on unequal sample sizes because it is unreasonable to assume that all studies in a meta-analysis will have the same (or even similar) sample sizes.

6.3 Level of Significance

Although some studies do not contain enough information to calculate an effect-size estimate, they may still provide information about the magnitude of the relation between the independent and dependent variables. Often, the information is in the form of a report of the decision yielded by a significance test (for example, a significant positive mean difference) or in the form of the direction of the effect without regard to its statistical significance (for example, a positive mean difference). The first of these corresponds to whether the test statistic exceeds a conventional critical value at a given significance level, such as $\alpha = .05$.[1] The second corresponds to whether the test statistic exceeds the rather unconventional level $\alpha=.5$. For the vote-counting procedures described in this chapter, we consider both the $\alpha=.05$ and $\alpha=.5$ significance levels.

6.4 Vote-Counting Situations

This chapter discusses two types of vote-counting situations. The first situation considers a collection of k independent studies in which each study compares two groups: an experimental group and a control group. The meta-analyst is interested in determining whether the group means differ and in estimating the magnitude of the mean difference across studies. The second situation considers a collection of k independent studies in which each study provides a correlation coefficient between

---

[1]Of course, you can use other values of α to define a significant positive result (for example, .01 or .10).

variables that have a bivariate normal distribution. The meta-analyst is interested in determining whether the correlations differ from zero and in estimating the magnitude of the correlation coefficient across studies.

In both situations, let $T_1, ..., T_k$ be independent estimators of population effect sizes $\theta_1, ..., \theta_k$ from k studies. Suppose that these k population effect sizes are homogeneous—that is, $\theta_1 = \theta_2 = \cdots = \theta_k = \theta$. The objective is to obtain an estimate and a confidence interval for θ using the k estimators $T_1, ..., T_k$. If the values of $T_1, ..., T_k$ are observed, you should use the methods discussed in Chapter 5 to estimate θ. The essential feature of vote-counting procedures is that the values of $T_1, ..., T_k$ are not observed. Even if the values of $T_1, ..., T_k$ are not observed, you can still estimate θ if the studies contain information about the direction and/or statistical significance of result. To estimate θ, simply count the number of times T_i exceeded the one-sided critical value $C_{\alpha, \, v_i}$, where v_i are the degrees of freedom for the ith study and α is the significance level. Specifically, if the significance level is $\alpha = .5$ (that is, $C_{\alpha = 0.5, v_i} = 0$ for all studies), count the number of studies with positive results. The proportion of positive results should be about .5 if θ is close to zero. If the significance level is $\alpha = .05$, count the number of studies with significant positive results. The proportion of significant positive results should be about .05 if θ is close to zero. If you also assume that the effect sizes are non-negative, the null and alternative hypotheses are

$$
\begin{aligned}
H_0: \quad & \theta_1 = \theta_2 = \cdots = \theta_k = \theta = 0 \\
H_A: \quad & \theta_1 = \theta_2 = \cdots = \theta_k = \theta > 0
\end{aligned}
\tag{6.1}
$$

For purposes of illustration, it is convenient to define an indicator variable V_i such that,

$$
V_i = \begin{cases} 1 & \text{if } T_i > C_{\alpha, \, v_i} \\ 0 & \text{if } T_i \le C_{\alpha, \, v_i} \end{cases}
\tag{6.2}
$$

Let p_i be the probability of $V_i = 1$. If you count the number of positive results, the probability is

$$p_i = \Pr(V_i = 1) = \Pr\left(T_i > C_{\alpha=.05, v_i}\right). \tag{6.3}$$

If you count the number of significant positive results, the probability is

$$p_i = \Pr(V_i = 1) = \Pr\left(T_i > C_{\alpha=.05, v_i}\right). \tag{6.4}$$

It has been shown that if T_i is approximately normally distributed with mean θ and variance σ_θ^2, then

$$p_i = \Pr\left(T_i > C_{\alpha, v_i}\right) \tag{6.5}$$

$$= \Pr\left[(T_i - \theta)/\sigma_\theta > \left(C_{\alpha, v_i} - \theta\right)/\sigma_\theta\right]$$

$$= 1 - \Phi\left[\left(C_{\alpha, v_i} - \theta\right)/\sigma_\theta\right],$$

where $\Phi(.)$ is the standard normal cumulative distribution function (Hedges & Olkin, 1985, pp. 85–91, 235). Solving Equation 6.5 for θ gives

$$\theta = C_{\alpha, v_i} - \sigma_\theta \Phi^{-1}(1 - p_i). \tag{6.6}$$

You can use the method of maximum likelihood to estimate θ by observing whether $T_i > C_{\alpha, v_i}$ in each of the k studies. The log-likelihood function is given by

$$L(\theta) = \sum_{i=1}^{k} \left[v_i \ln(p_i)\right] + \left[(1 - v_i)\ln(1 - p_i)\right], \tag{6.7}$$

where $\ln(.)$ is the natural logarithm function and v_i is the observed value of V_i. The log-likelihood function $L(\theta)$ can be maximized over θ to obtain the maximum likelihood estimator (MLE) $\hat{\theta}$. Generally, however, there is no closed form expression for the MLE $\hat{\theta}$, and $\hat{\theta}$ must be obtained numerically. This chapter contains the SAS code for obtaining MLEs for the two measures of effect that dominate the meta-analytic literature: the standardized mean difference and the correlation coefficient (see Appendices 6.2 and 6.3, respectively).

When the number of studies combined is large, $\hat{\theta}$ is approximately normally distributed, and the large-sample variance of $\hat{\theta}$ is given by

$$\text{Var}\left(\hat{\theta}\right) = \left\{ \sum_{i=1}^{k} \frac{D_i^{(1)^2}}{p_i\left(1 - p_i\right)} \right\}^{-1}, \tag{6.8}$$

where $D_i^{(1)}$ is the first order derivative of p_i evaluated at $\theta = \hat{\theta}$ (Hedges & Olkin, 1985, p. 70). The upper and lower bounds of a $100(1 - \alpha)\%$ confidence interval for the population effect size θ are given by

$$\hat{\theta} - Z_{\alpha/2}\sqrt{\text{Var}\left(\hat{\theta}\right)} \leq \theta \leq \hat{\theta} + Z_{\alpha/2}\sqrt{\text{Var}\left(\hat{\theta}\right)}, \tag{6.9}$$

where $Z_{\alpha/2}$ is the two-sided critical value from the standard normal distribution at significance level α.

Section 6.4.1 illustrates how to obtain an estimate and a confidence interval for the population standardized mean difference δ. Section 6.4.2 illustrates how to obtain an estimate and a confidence interval for the population correlation coefficient ρ.

6.4.1 Vote-Counting Procedures for Estimating the Population Standardized Mean Difference

When the primary studies in question compare two groups, either through experimental (treatment) versus control group comparisons or through orthogonal contrasts, the effect-size estimate often is expressed as some form of standardized difference between the group means. Suppose that the data arise from a series of k independent studies, each of which compares a treatment or experimental group (E) with a control group (C). Let $Y_{E_{ij}}$ and $Y_{C_{ij}}$ denote the jth observations in the experimental and control groups of the ith study. Define the population standardized mean difference as

$$\delta_i = \frac{\mu_{E_i} - \mu_{C_i}}{\sigma_i}, \tag{6.10}$$

where μ_{E_i} and μ_{C_i} are the population means for the experimental and control groups in the ith study, respectively, and σ_i is the common population standard deviation in the ith study. Thus, you substitute δ_i for the more general parameters θ_i in Equation 6.1. The sample estimator of δ_i is

$$g_i = \frac{\overline{Y}_{E_i} - \overline{Y}_{C_i}}{S_{POOLED_i}}, \tag{6.11}$$

where $\overline{Y}_{E_i}$ and $\overline{Y}_{C_i}$ are the sample means for the experimental and control groups in the ith study, respectively, and S_{POOLED_i} is the pooled sample standard deviation in the ith study (Hedges, 1981).[2] Thus, you substitute g_i for the more general estimator T_i in Equations 6.2, 6.3, 6.4, and 6.5.

Example 6.1 *Obtaining an estimate and a confidence interval for the population standardized mean difference based on the proportion of positive results*

The data set for Example 6.1 was taken from previous meta-analytic studies of alcohol and aggression (see Bushman, 1993; Bushman & Cooper, 1990). One explanation of intoxicated aggression is based on alcohol-related expectancies – people expect alcohol to increase aggression. In laboratory experiments, this expectancy interpretation of intoxicated aggression can be tested by comparing the level of aggression for participants who expect an alcoholic beverage but receive a nonalcoholic beverage (that is, placebo) with the level of aggression for participants who expect and receive a nonalcoholic beverage (that is, control). The placebo versus control comparison gives the pure effects of alcohol-related expectancies on aggression uncontaminated by the pharmacological effects of alcohol. In the laboratory, aggression is measured by having participants give noxious physical (for example, electric shocks, noise blasts) or verbal (for example, hostile comments, negative evaluations) stimuli to another person or by having participants take away positive stimuli (for example, money) from another person.

---

[2]Alternatively, you can use the unbiased estimator g_{U_i} (see Chapter 5, Section 5.2.1). With vote-counting methods, whether you use g_i or g_{U_i} makes little difference because the effect-size estimates are only used to obtain an initial estimate for the iterative numerical procedure.

Table 6.1 lists the results from 13 laboratory experiments that included placebo
and control groups; all participants were male social drinkers (Bushman &
Wang, 1996, Table 1). Sample standardized mean differences were calculated for
nine of the 13 studies (that is, studies 1–9) included in Table 6.1. The remaining
four studies (that is, studies 10–13) did not contain enough information to calculate
standardized mean differences but did contain information about the direction of
results. In practice, you should not use vote-counting procedures on studies that
include standardized mean differences because effect-size procedures are more
accurate (see Chapter 5 for a discussion of effect-size procedures). We illustrate
vote-counting procedures here because in the next chapter we describe a procedure
for combining effect-size estimates and vote counts.

Table 6.1 *Studies testing the effects of alcohol-related expectancies on aggression
(Bushman & Wang, 1996)*

Study	n_E	n_C	g	p	Direction
1	19	19	1.028	< .05	+
2	24	24	0.903	< .05	+
3	24	24	0.552	< .05	+
4	24	24	0.328	ns	+
5	12	12	0.193	ns	+
6	24	24	-0.036	ns	-
7	24	24	-0.332	ns	-
8	15	15	-0.540	ns	-
9	12	12	-0.660	ns	-
10	16	16		ns	+
11	12	12		ns	+
12	10	10		ns	+
13	10	10		ns	-

Note: n_E = number of participants in placebo group. n_C = number of participants in
control group. g = Hedges' estimate for the population standardized mean difference. *ns* =
nonsignificant at the .05 level.

You can call the SAS macro VOTERUN in Appendix 6.4 to analyze the data in Table 6.1. Use the following command to access the SAS macro VOTERUN:

%**VOTERUN**(*indata,index,alpha,level*);

where

indata is the name of the SAS data set. The SAS data set name for Example 6.1 is EX61. The data set EX61 contains four variables: NE, NC, SIG, and EFFECT. NE and NC are the experimental and control group sample sizes (that is, column n_E and n_C in Table 6.1), respectively. SIG is the indicator variable for the Direction column in Table 6.1. SIG is coded 1 if the direction is positive, and 0 otherwise. EFFECT is the value of any effect-size estimate when available (that is, Glass's $\hat{\Delta}$, Cohen's d, Hedges' g, or Hedges' g_U); when unavailable, a period (.) is used to denote a missing estimate. In Example 6.1, Hedges' g was used (that is, column g in Table 6.1). The variable EFFECT is only used to obtain an initial estimate for the iterative numerical procedure. The initial estimate is set to zero if all of the studies have missing effect-size estimates.

index is the effect-size index. Set *index* to MEAN if the effect-size index is the standardized mean difference. Set *index* to CORR if the effect-size index is the Pearson product-moment correlation coefficient. In Example 6.1, *index* was set to MEAN because we want to estimate the standardized mean difference in aggression between the placebo and control groups.

alpha is the significance level for the results being counted. If you are counting the number of positive results, set *alpha* to .50. If you are counting the number of significant positive results, set *alpha* to .05. In Example 6.1, *alpha* is set to .50.

level is the level of confidence for the confidence interval (for example, .95 for a 95% confidence interval).

The following SAS code was used to obtain 90%, 95%, and 99% confidence intervals for the population standardized mean difference from the proportion of positive results in Table 6.1. The results are printed in Output 6.1.

```
options nocenter nodate pagesize=54 linesize=80 pageno=1;
libname aa "d:\metabook\ch6\dataset";
%voterun(aa.ex61,"Mean",0.5,0.90);
data result;
    set voteaout;
%voterun(aa.ex61,"Mean",0.5,0.95);
data result;
    set result voteaout;
%voterun(aa.ex61,"Mean",0.5,0.99);
data result;
    set result voteaout;
proc print data=result round noobs;
    format level 3.2 delta vdelta lower upper 5.3;
    title;
run;
```

Output 6.1 *Confidence intervals for the population standardized mean difference based on the proportion of positive results in Table 6.1*

LEVEL	DELTA	VDELTA	LOWER	UPPER
.90	0.102	0.014	-.095	0.300
.95	0.102	0.014	-.133	0.338
.99	0.102	0.014	-.207	0.412

In Output 6.1, LEVEL is the level of confidence for the confidence interval, DELTA is the estimate for the population standardized mean difference, VDELTA is the variance of the estimate, and LOWER and UPPER are the respective lower and upper bounds of the confidence interval. For example, the estimate of the population standardized mean difference is 0.102 with a 95% confidence interval ranging from -0.133 to 0.338. The confidence interval includes the value zero.

Example 6.2 *Obtaining an estimate and a confidence interval for the population standardized mean difference based on the proportion of significant positive results*

The SAS macro VOTERUN (see Appendix 6.4) estimates the population standardized mean difference from the proportion of *significant* positive results. However, the definition of the variables SIG and ALPHA are different. Because you are counting positive significant results, the value of SIG is 1 when the study's result is positive and significant at the .05 level, and is 0 otherwise. In addition, the value of ALPHA is set to .05. The data set for Example 6.2 is called EX62.

The following SAS code was used to obtain 90%, 95%, and 99% confidence intervals for the population standardized mean difference from the proportion of *significant* positive results in Table 6.1. The results are printed in Output 6.2.

```
%voterun(aa.ex62,"Mean",0.05,0.90);
data result;
    set voteaout;
%voterun(aa.ex62,"Mean",0.05,0.95);
data result;
    set result voteaout;
%voterun(aa.ex62,"Mean",0.05,0.99);
data result;
    set result voteaout;
proc print data=result round noobs;
    format level 3.2 delta vdelta lower upper 5.3;
    title;
run;
```

Output 6.2 *Confidence intervals for the population standardized mean difference based on the proportion of significant positive results in Table 6.1*

```
LEVEL     DELTA     VDELTA     LOWER     UPPER

  .90     0.341     0.021      0.103     0.579
  .95     0.341     0.021      0.057     0.625
  .99     0.341     0.021     -.032      0.714
```

As you can see in Output 6.2, the estimate of the population standardized mean difference is 0.341 with a 95% confidence interval ranging from 0.057 to 0.625. The confidence interval does not include the value zero.

You can use the SAS macro VOTERUN to obtain an estimate and a confidence interval for the population standardized mean difference at any significance level. If you are counting the number of positive results that are significant at the .01 level (for example, SIG equals 1 when the study result is positive and significant at the .01 level, and 0 otherwise) and ALPHA is set to .01.

6.4.2 Vote-Counting Procedures for Estimating the Population Correlation Coefficient

Suppose that the data arise from a series of k independent studies, each of which measures the linear relation between two continuous variables, say X and Y. The population correlation coefficient for the ith study is

$$\rho_i = \frac{\text{Cov}(X_i, Y_i)}{\sigma_{X_i} \sigma_{Y_i}} , \qquad (6.13)$$

where $\text{Cov}(X_i, Y_i)$ is the population covariance of X and Y in the ith study, σ_{X_i} and σ_{Y_i} are the population standard deviations for X and Y, respectively, in the ith study. Thus, you substitute ρ_i for the more general parameter θ_i in Equation 6.1. The sample estimator of ρ_i is the Pearson product-moment correlation coefficient:

$$r_i = \frac{\sum_{j=1}^{n_i}\left(X_{ij} - \bar{X}_i\right)\left(Y_{ij} - \bar{Y}_i\right)}{\sqrt{\left[\sum_{j=1}^{n_i}\left(X_{ij} - \bar{X}_i\right)^2\right]\left[\sum_{j=1}^{n_i}\left(Y_{ij} - \bar{Y}_i\right)^2\right]}}, \tag{6.14}$$

where $\bar{X}_i$ and $\bar{Y}_i$ are the respective mean values for the independent and dependent variables in the ith study. Thus, you substitute r_i for the more general estimator T_i in Equations 6.2, 6.3, 6.4, and 6.5.

Example 6.3 *Obtaining an estimate and a confidence interval for the population correlation coefficient based on the proportion of positive results*

The data set for Example 6.3 comes from a meta-analysis by Gerrard, Gibbons, and Bushman (1996). The studies in this meta-analysis tested the hypothesis that perceived risk to human immunodeficiency virus (HIV) infection motivates preventative sexual behaviors (for example, using condoms). To measure perceived risk to HIV infection, all researchers used some variation of the question "What is the likelihood that you will contract HIV?" or "What is the likelihood that you will develop acquired immunodeficiency syndrome (AIDS)? "

Table 6.2 contains the results from 36 independent samples of participants. Pearson product-moment correlation coefficients were calculated for 12 of the 36 samples (that is, samples 1–12). The remaining 24 samples (that is, samples 13–36) did not contain enough information to calculate Pearson product-moment correlation coefficients, but did contain information about the direction and/or significance of the results. In practice, vote-counting procedures should not be used on studies that include correlations because effect-size procedures are more accurate (see Chapter 5 for a discussion of effect-size procedures). We illustrate vote-counting procedures here because in the next chapter we describe a procedure for combining effect-size estimates and vote counts.

Table 6.2 *Studies correlating perceived vulnerability to HIV infection with precautionary sexual behavior*

Study	N	r	p	Direction
1	294	.240	<.05	+
2	60	.240	ns	+
3	578	.123	<.05	+
4	459	.069	ns	+
5	241	-.050	ns	-
6	866	-.062	ns	-
7	493	-.130	<.05	-
8	212	-.162	<.05	-
9	272	-.210	<.05	-
10	147	-.260	<.05	-
11	481	-.342	<.05	-
12	99	-.346	<.05	-
13	1,127		<.05	+
14	312		<.05	+
15	99		ns	+
16	432		ns	+
17	637		ns	+
18	909		ns	+
19	580		ns	-
20	544		ns	-
21	501		ns	-
22	402		ns	-
23	266		ns	-
24	205		ns	-
25	197		ns	-
26	195		ns	-
27	152		ns	-
28	105		ns	-
29	111		ns	-
30	114		<.05	-
31	129		<.05	-
32	206		<.05	-
33	466		<.05	-
34	513		<.05	-
35	602		<.05	-
36	964		<.05	-

Note: ns = nonsignificant at the .05 level; + = positive result (that is, in predicted direction); - = negative result (that is, in opposite direction). N = total sample size. r = Pearson product moment correlation coefficient.

The SAS macro VOTERUN(*indata,index,alpha,level*) in Appendix 6.4 was used to obtain an estimate and confidence interval for the population correlation coefficient from the proportion of positive results in Table 6.2. The macro parameters are the same as those given for Example 6.1, with the following exceptions. The input data set *indata*, called EX63 for this example, contains three variables: NN, SIG, and RR. NN is the total sample size (column *N* in Table 6.2). SIG is the indicator variable for the Direction column in Table 6.2. RR is the correlation coefficient for the study, when available; when unavailable, a period (.) is used to denote a missing estimate. RR is only used to obtain an initial estimate for the iterative numerical procedure. The initial estimate is set to zero if all of the studies have missing correlations. In Example 6.3, *index* is set to CORR because the effect-size index is the Pearson product-moment correlation coefficient.

The following SAS code was used to obtain 90%, 95%, and 99% confidence intervals for the population correlation coefficient from the proportion of positive results in Table 6.2. The results are printed in Output 6.3.

```
options nocenter nodate pagesize=54 linesize=80 pageno=1;
libname aa "a:\metabook\ch6\dataset";
%voterun(aa.ex63,"corr",0.5,0.90);
data result;
    set votebout;
%voterun(aa.ex63,"corr",0.5,0.95);
data result;
    set result votebout;
%voterun(aa.ex63,"corr",0.5,0.99);
data result;
    set result votebout;
proc print data=result round noobs;
    format level 3.2 rho vrho lower upper 6.4;
run;
```

Output 6.3 *Confidence intervals for the population correlation coefficient based on the proportion of positive results in Table 6.2*

LEVEL	RHO	VRHO	LOWER	UPPER
.90	-.0233	0.0001	-.0417	-.0049
.95	-.0233	0.0001	-.0452	-.0013
.99	-.0233	0.0001	-.0521	0.0055

In Output 6.3, LEVEL is the level of confidence for the confidence interval, RHO is the estimate for the population correlation coefficient, VRHO is the variance of the estimate, and LOWER and UPPER are the respective lower and upper bounds of the confidence interval. The estimated population correlation is −0.023 with a 95% confidence interval ranging from −0.0452 to −0.0013. The confidence interval does not include the value zero.

Example 6.4 *Obtaining an estimate and a confidence interval for the population correlation coefficient based on the proportion of significant positive results*

The SAS macro VOTERUN (*indata,index,alpha,level*) in Appendix 6.4 was used to estimate the population correlation from the proportion of *significant* positive results in Table 6.2. Because you are counting significant positive results, the value for ALPHA is .05. In addition, SIG is assigned the value 1 if the sample correlation for the study is positive and significant at the .05 level, and 0 otherwise. The SAS data set name for Example 6.4 is EX64. The results from the following SAS code are printed in Output 6.4:

```
options nocenter nodate pagesize=54 linesize=80 pageno=1;
libname aa "d:\metabook\ch6\dataset";
%voterun(aa.ex64,"corr",0.05,0.90);
data result;
    set votebout;
%voterun(aa.ex64,"corr",0.05,0.95);
data result;
    set result votebout;
%voterun(aa.ex64,"corr",0.05,0.99);
```

```
data result;
   set result votebout;
proc print data=result round noobs;
   format level 3.2 rho vrho lower upper 6.4;
run;
```

Output 6.4 *Confidence intervals for the population correlation coefficient based on the proportion of significant positive results in Table 6.2*

LEVEL	RHO	VRHO	LOWER	UPPER
.90	-.0473	0.0002	-.0679	-.0266
.95	-.0473	0.0002	-.0719	-.0227
.99	-.0473	0.0002	-.0796	-.0149

As you can see in Output 6.4, the estimated population correlation is −0.047 with a 95% confidence interval ranging from −0.072 to −0.023. The confidence interval does not include the value zero.

6.5 Conclusions

Missing effect-size estimates pose a particularly difficult problem in meta-analysis. Often, research reports do not include enough information to permit the calculation of an effect-size estimate. Even if research reports do not contain enough information to permit the calculation of effect-size estimates, they may still provide information about the magnitude of the effect size. Often, the information is in the form of a report of the decision that is yielded by the significance test (for example, a significant positive mean difference) or in the form of a direction without regard to its statistical significance (for example, a positive mean difference). Vote-counting procedures use the information about the direction and statistical significance of study results to obtain an estimate and a confidence interval for a population effect-size.

One limitation of vote-counting procedures is that they cannot be used if all of the results are in the same direction. The method of maximum likelihood cannot be used if all results are in the same direction because there is not a unique corresponding maximum likelihood estimator. If all of the results are in the same direction, you can use a Bayes estimate instead of a vote-count estimate (see Hedges & Olkin, 1985, p. 300).

Vote-counting procedures are not very useful when they are used alone (unless none of the studies in the meta-analysis report effect-size estimates). Vote-counting procedures are quite useful, however, when they are used in conjunction with effect-size procedures. In the next chapter, we discuss how to combine sample effect sizes and vote counts to estimate population effect sizes.

6.6 References

Bushman, B. J. (1993). Human aggression while under the influence of alcohol and other drugs: An integrative research review. *Current Directions in Psychological Science, 2,* 148–152.

Bushman, B. J. (1994). Vote-counting procedures in meta-analysis. In H. Cooper & L. V. Hedges (Eds.), *The handbook of research synthesis* (pp. 193–213). New York: Russell Sage Foundation.

Bushman, B. J., & Cooper, H. M. (1990). Effects of alcohol on human aggression: An integrative research review. *Psychological Bulletin, 107,* 341–354.

Bushman, B. J., & Wang, M. C. (1995). A procedure for combining sample correlation coefficients and vote counts to obtain an estimate and a confidence interval for the population correlation coefficient. *Psychological Bulletin, 117,* 530–546.

Bushman, B. J., & Wang, M. C. (1996). A procedure for combining sample standardized mean differences and vote counts to estimate the population standardized mean difference in fixed effects models. *Psychological Methods, 1,* 66–80.

Gerrard, M., Gibbons, F. X., & Bushman, B. J. (1996). Relation between perceived vulnerability to HIV and precautionary sexual behavior. *Psychological Bulletin, 119,* 390–409.

Hedges, L. V. (1981). Distribution theory for Glass's estimator of effect size and related estimators. *Journal of Educational Statistics, 6,* 107–128.

Hedges, L. V. (1982). Estimation of effect size from a series of independent experiments. *Psychological Bulletin, 92,* 490–499.

Hedges, L. V. (1986). *Estimating effect sizes from vote counts or box score data.* Paper presented at the annual meeting of the American Educational Research Association, San Francisco, CA.

Hedges, L. V., & Olkin, I. (1980). Vote-counting methods in research synthesis. *Psychological Bulletin, 88,* 359–369.

Hedges, L. V., & Olkin, I. (1985). *Statistical methods for meta-analysis.* New York: Academic Press.

Light, R. J., & Smith, P. V. (1971). Accumulating evidence: Procedures for resolving contradictions among different research studies. *Harvard Educational Review, 41,* 429–471.

6.7 Appendices

Appendix 6.1: SAS/IML Module for Obtaining the Probability of the Vote-Counting Estimate Using the Large Sample Approximation Method

```
libname aa "d:\metabook\ch6\dataset";
proc iml;
/**********************************************************/
/*  Compute the probability of the vote-counting estimate */
/*  with the large sample approximation method            */
/**********************************************************/
   start votecdf1(alpha,size,rho,votecdf);
   if (alpha = 0.5) then do;
      tem = -sqrt(size)*rho/(1-rho*rho);
      if (tem > 7.941) then tem=7.941;
      else if (tem < -8.222) then tem=-8.222;
      votecdf = 1 - probnorm(tem);
   end;
   else do;
      malpha = alpha * 2;
      cst = 0.5*size-1;
      can = sqrt(1 - betainv(malpha,cst,0.5));
      tem = (can-sqrt(size)*rho)/(1-rho*rho);
      if (tem > 7.941) then tem=7.941;
      else if (tem < -8.222) then tem=-8.222;
      votecdf = 1 - probnorm(tem);
   end;
   finish votecdf1;
   reset storage=aa.imlrout;
   store module=(votecdf1);
run;
quit;
```

Appendix 6.2: SAS/IML Module for Obtaining a Confidence Interval for the Population Standardized Mean Difference Using Vote-Counting Procedures

```
libname aa "d:\metabook\ch6\dataset";
/**********************************************************/
/* Purpose:                                            */
/* Vote-Counting Methods for Obtaining a Confidence    */
/* Interval for the Population Standardized Mean        */
/* Difference                                          */
/* Input Information:                                  */
/* NE: The experimental group sample size              */
/* NC: The control group sample size                   */
/* SIG: SIG=1 if the effect-size estimate for the study */
/*      is significantly different from zero at         */
/*      level = ALPHA                                   */
/*      SIG=0 otherwise                                 */
/* EFFECT: Any effect-size estimate for studies that   */
/*         provide enough information to compute it.    */
/*      For studies with missing effect-size estimates, */
/*      EFFECT = . (for a missing value).              */
/* ALPHA: The level of significance for each study     */
/*         under review                               */
/* LEVEL: The confidence level. of the confidence      */
/*         interval                                    */
/* Output Information:                                 */
/*    Output File Name: VOTEAOUT                        */
/*    Variable Name:                                   */
/*    DELTA: Effect-size estimate based on             */
/*           vote-counting methods                      */
/*    VDELTA: Corresponding variance for the           */
/*            effect-size estimate                     */
/*    LOWER: Lower bound for the confidence interval   */
/*    UPPER: Upper bound for the confidence interval   */
/*    LEVEL: The confidence level. of the confidence   */
/*           interval                                  */
/**********************************************************/
proc iml;
   start votea(ne,nc,sig,effect,alpha,level);
/**********************************************************/
/*   Initialize some constants                         */
/*   DF: Degrees of freedom for each study             */
/*   SVC: Square root of the harmonic mean for the     */
/*        sample sizes                                 */
/*   TT: Upper quantile for each study                 */
/**********************************************************/
   df=ne+nc-2;
   svc=sqrt(ne#nc/(ne+nc));
   tt=j(nrow(ne),1,0);
   if (alpha < 0.5) then do;
      tt=betainv(0.9,0.5,0.5#df);
      tt=sqrt((tt#df)/(1-tt));
   end;
```

```
/***********************************************************/
/*   Start the iteration to find the effect-size          */
/*        estimate                                         */
/*   DELTA: Effect-size estimate based on vote-count       */
/*   VDELTA: Variance for the effect-size estimate         */
/***********************************************************/
   again=j(5,1,0);
   again[5,1]=max(effect);
   again[1,1]=min(effect);
   again[3,1]=0.5*(again[1,1]+again[5,1]);
   again[2,1]=0.5*(again[1,1]+again[3,1]);
   again[4,1]=0.5*(again[3,1]+again[5,1]);
   abserr=1;
   k=1;
   xx=j(5,1,0);
   yy=j(nrow(ne),1,0);
   do while (abserr > 2d-5 & k <= 30);
      large = 1;
      do I = 1 to 5;
         xx[i,1] = 0;
      do j = 1 to nrow(ne);
         delta = again[i,1] * svc[j,1];
         res = 1 - probt(tt[j,1],df[j,1],delta);
         xx[i,1] = xx[i,1] + (sig[j,1]*log(res)+
                   (1-sig[j,1])*log(1-res));
      end;
      if (i > 1) then do;
         if (xx[i,1] > xx[i-1,1]) then large=i;
      end;
   end;
   again[1,1] = again[large-1,1];
   again[5,1] = again[large+1,1];
   again[3,1] = again[large,1];
   again[4,1] = 0.5 * (again[3,1]+again[5,1]);
   again[2,1] = 0.5 * (again[1,1]+again[3,1]);
   k=k+1;
   abserr=abs(again[5,1]-again[1,1]);
   end;
DELTA=again[3,1];
```

```
/**********************************************************/
/*  Find the maximum likelihood estimate for the          */
/*  variance of the vote-count estimate for the           */
/*  effect-size                                           */
/**********************************************************/
vdelta=0;
do I = 1 to nrow(ne);
   res=svc[I,1]*delta;
   tt[i,1]=probnorm(res);
   yy[i,1]=(svc[i,1]/sqrt(8*atan(1)))*exp(res*res*(-0.5));
   vdelta=vdelta+yy[i,1]*yy[i,1]/(tt[i,1]*(1-tt[i,1]));
   end;
vdelta=1/vdelta;
/**********************************************************/
/*  Compute the confidence interval for the population    */
/*  standardized mean difference                          */
/*  LOWER: Lower bound for confidence interval            */
/*  UPPER: Upper bound for confidence interval            */
/*  LEVEL: The confidence level of the confidence         */
/*         interval                                       */
/*  DELTA: Effect-size estimate based on vote-count       */
/*  VDELTA: Variance for the effect-size estimate         */
/**********************************************************/
   estimate=delta;
   varest=vdelta;
   delta=j(nrow(level),1,0);
   vdelta=j(nrow(level),1,0);
   lower=j(nrow(level),1,0);
   upper=j(nrow(level),1,0);
   do i = 1 to nrow(level);
      lbound=estimate
             +probit(0.5*(1-level[i,1]))*sqrt(varest);
      ubound=estimate
             +probit(0.5*(1+level[i,1]))*sqrt(varest);
      delta[i,1]=estimate;
      vdelta[i,1]=varest;
      lower[i,1]=lbound;
      upper[i,1]=ubound;
   end;
   create voteaout var{level delta vdelta lower upper};
   append;
   close voteaout;
   finish votea;
   reset storage=aa.imlrout;
   store module=(votea);
 run;
 quit;
```

Appendix 6.3: SAS Macro for Obtaining a Confidence Interval for the Population Correlation Coefficient Using Vote-Counting Procedures

```
libname aa "d:\metabook\ch6\dataset";
/**********************************************************/
/* Purpose: Vote-Counting Methods for Obtaining a         */
/*          Confidence Interval for the Pearson Product   */
/*          Correlation Coefficient                       */
/*                                                        */
/* Input Information:                                     */
/* NN: Total sample size                                  */
/* SIG: SIG = 1 if the correlation for the study is       */
/*              significantly different from zero at      */
/*              level = ALPHA                             */
/*      SIG = 0 otherwise                                 */
/* RR: Correlation for studies that provided this         */
/*     information.  RR sets to the maximum value for     */
/*     half of the studies that did not provide this      */
/*     information                                        */
/*     RR sets to minimum value for another half of the   */
/*     studies that did not provide this information.     */
/* ALPHA: The level of significance for each study        */
/*        under review                                    */
/* LEVEL: The confidence level for the combined           */
/*        confidence interval                             */
/*                                                        */
/* Output Information:                                    */
/* Print Output:                                          */
/* (1) RHO: Effect-size estimate based on vote-count      */
/* (2) VRHO: Variance for the effect-size estimate        */
/* (3) LOWER: Lower bound for confidence interval         */
/* (4) UPPER: Upper bound for confidence interval         */
/* (5) LEVEL: Confidence level for the combined           */
/*            interval                                    */
/**********************************************************/
proc iml;
   start voteb(nn,sig,rr,alpha,level);
/**********************************************************/
/*  Start the iteration to find the effect-size          */
/*  estimate                                             */
/*  RHO: Effect-size estimate based on vote-counting      */
/*       method                                           */
/*  VRHO: Variance for the effect-size estimate          */
/**********************************************************/
   again=j(5,1,0);
   again[5,1]=max(rr);
   again[1,1]=min(rr);
   again[3,1]=0.5*(again[1,1]+again[5,1]);
   again[2,1]=0.5*(again[1,1]+again[3,1]);
   again[4,1]=0.5*(again[3,1]+again[5,1]);
   abserr=1;
   kk=1;
```

```
xx=j(5,1,0);
dd=j(nrow(nn),1,0);
pp=j(nrow(nn),1,0);
do while ((abserr > 2d-5) & (kk < 50));
    large = 1;
    do i = 1 to 5;
        rho=again[i,1];
        xx[i,1] = 0;
        do j = 1 to nrow(nn);
            size=nn[j,1];
            run votecdf1(alpha,size,rho,votecdf);
            xx[i,1]=xx[i,1]+sig[j,1]*log(votecdf)+
                (1-sig[j,1])*(log(1-votecdf));
        end;
        if (i > 1) then do;
            if (xx[i,1] > xx[i-1,1]) then large = I;
        end;
    end;
    again[1,1] = again[large-1,1];
    again[5,1] = again[large+1,1];
    again[3,1] = 0.5*(again[1,1]+again[5,1]);
    again[2,1] = 0.5*(again[1,1]+again[3,1]);
    again[4,1] = 0.5*(again[3,1]+again[5,1]);
    kk=kk+1;
    abserr=abs(again[5,1]-again[1,1]);
end;
/***********************************************************/
/*  Compute the confidence interval for the Pearson       */
/*       Product correlation:                             */
/*  LOWER: Lower bound for confidence interval            */
/*  UPPER: Upper bound for confidence interval            */
/*  LEVEL: Confidence level for the combined interval     */
/*  RHO: Effect-size estimate                             */
/*  VRHO: Variance for the effect-size estimate           */
/***********************************************************/
esrho=again[3,1];
vesrho=0;
tt3=8*atan(1);
rho=esrho;
```

```
do i = 1 to nrow(nn);
   size=nn[I,1];
   run votecdf1(alpha,size,rho,votecdf);
   pp[i,1]=votecdf;
   tt1=(1+esrho*esrho)/(1-esrho*esrho)##2;
   tt2=exp(-0.5*(nn[i,1]*esrho*esrho)/(1-esrho*esrho)**2);
   dd[i,1]=sqrt(nn[i,1]/tt3)*tt1*tt2;
   vesrho=vesrho+dd[i,1]*dd[i,1]/(pp[i,1]*(1-pp[i,1]));
   end;
   vesrho=1/vesrho;
   estimate=esrho;
   varest=vesrho;
   rho=j(nrow(level),1,0);
   vrho=j(nrow(level),1,0);
   lower = estimate + probit((1-level)*0.5)*sqrt(varest);
   upper = estimate + probit((1+level)*0.5)*sqrt(varest);
   do i = 1 to nrow(level);
      rho[i,1]=estimate;
      vrho[I,1]=varest;
   end;
   create votebout var{LEVEL RHO VRHO LOWER UPPER};
   append;
   close votebout;
   finish voteb;
   reset storage=aa.imlrout;
   store module=(voteb);
run;
quit;
```

Appendix 6.4: SAS Macro for Estimating Population Effect Sizes Using Vote-Counting Procedures

```
%macro voterun(indata,index,alpha,level);
proc iml;
   reset nolog;
   reset storage=ch6.imlrout;
   alpha=&alpha;
   level=&level;
   if (upcase(&index)="MEAN") then do;
      load module=(votea);
      use &indata;
      read all;
      run votea(ne,nc,sig,effect,alpha,level);
      end;
   else if (upcase(&index)="CORR") then do;
      load module=(voteb votecdf1);
      use &indata;
      read all;
      run voteb(nn,sig,rr,alpha,level);
      end;
run;
quit;
%mend voterun;
```

chapter 7

Combining Effect-Size Estimates and Vote Counts

7.1 Introduction

Missing data is perhaps the largest problem facing the practicing meta-analyst. Missing effect-size estimates poses a particularly difficult problem because meta-analytic procedures cannot be used at all without a statistical measure for the results of a study (Pigott, 1994). Sometimes, research reports do not include enough information (for example, means, standard deviations, statistical tests) to permit the calculation of an effect-size estimate. Unfortunately, the proportion of studies with missing effect-size estimates in a research synthesis is often quite large, about 25% in psychological studies (Bushman & Wang, 1995, 1996).

Currently, the most common "solutions" to the problem of missing effect-size estimates are to (a) omit from the review those studies with missing effect-size estimates and analyze only complete cases, (b) set the missing effect-size estimates equal to zero, (c) set the missing effect-size estimates equal to the mean obtained from studies with effect-size estimates, and (d) set studies equal to the conditional mean that is obtained from studies with effect-size estimates (that is, Buck's method, 1960. Unfortunately, all of these procedures have serious problems that limit their usefulness (Bushman & Wang, 1996). Imputing the value zero for missing effect-size estimates, for example, underestimates the population effect size and artificially reduces the variance of effect-size estimates.

We proposed an alternative procedure for handling missing effect-size estimates that overcomes the problems associated with conventional procedures for missing data (Bushman & Wang, 1996). Our procedure, called the combined procedure, combines sample effect-sizes and vote counts to estimate the population effect size. In this chapter, we describe how to use the combined procedure to obtain an estimate and a confidence interval for two common effect-size indexes: the standardized mean difference and the correlation coefficient. We assume fixed-effects models for both types of effect sizes (see Chapter 8 for a discussion of fixed-effects models). However, you can use the combined procedure for different effect-size indexes and for random-effects models.

7.2 Using the Combined Procedure to Estimate the Population Standardized Mean Difference

The combined procedure for standardized mean differences (Bushman & Wang, 1996) uses the effect-size and vote counting procedures that are described earlier in this book (see Chapters 5 and 6, respectively). Out of k independent studies, suppose that the first m_1 studies ($m_1 < k$) report enough information to compute standardized mean differences, that the next m_2 studies ($m_2 < k$) only report the statistical significance of results, and that the last m_3 studies ($m_3 < k$) only report the direction of results, where $m_1 + m_2 + m_3 = k$. Use effect-size procedures (see Chapter 5) to obtain Hedges' (1982) unbiased estimator g_{U_+} of the population standardized mean difference δ using the first m_1 studies. Use vote-counting procedures (see Chapter 6) to obtain a maximum likelihood estimator $\hat{\delta}_S$ of δ using the second m_2 studies. Also use vote-counting procedures to obtain the maximum likelihood estimator $\hat{\delta}_D$ of δ using the last m_3 studies. The respective variances for g_{U_+}, $\hat{\delta}_S$, and $\hat{\delta}_D$ are $\mathrm{Var}\left(g_{U_+}\right)$ (Equation 5.21, where $\hat{\delta}_i = g_{U_i}$), $\mathrm{Var}\left(\hat{\delta}_S\right)$ (Equation 6.8, where p_i is the proportion of significant positive mean differences), and $\mathrm{Var}\left(\hat{\delta}_D\right)$ (Equation 6.8, where p_i is the proportion of positive mean differences). The combined weighted estimator of δ is given by

$$\hat{\delta}_C = \frac{g_{U_+}\big/\mathrm{Var}\left(g_{U_+}\right)+\hat{\delta}_S\big/\mathrm{Var}\left(\hat{\delta}_S\right)+\hat{\delta}_D\big/\mathrm{Var}\left(\hat{\delta}_D\right)}{1\big/\mathrm{Var}\left(g_{U_+}\right)+1\big/\mathrm{Var}\left(\hat{\delta}_S\right)+1\big/\mathrm{Var}\left(\hat{\delta}_D\right)}. \tag{7.1}$$

Because g_{U_+}, $\hat{\delta}_S$, and $\hat{\delta}_D$ are consistent estimators of δ, the combined estimator $\hat{\delta}_C$ also will be consistent if (a) the studies that contain each type of information (that is, information used to calculate effect-size estimates, information about the direction and significance of results, information about only the direction of results) can be considered a random sample from the population of studies that contain the same type of information and (b) m_1, m_2, and m_3 are all sufficiently large and all converge to infinity at the same rate. The variance of the combined estimator is given by

$$\text{Var}\left(\hat{\delta}_C\right) = \frac{1}{1\big/\text{Var}\left(g_{U_+}\right) + 1\big/\text{Var}\left(\hat{\delta}_S\right) + 1\big/\text{Var}\left(\hat{\delta}_D\right)}. \tag{7.2}$$

Because $\dfrac{1}{\text{Var}\left(g_{U_+}\right)} > 0$, $\dfrac{1}{\text{Var}\left(\hat{\delta}_S\right)} > 0$, and $\dfrac{1}{\text{Var}\left(\hat{\delta}_D\right)} > 0$, the combined estimator is more efficient (that is, has smaller variance) than g_{U_+}, $\hat{\delta}_S$, and $\hat{\delta}_D$.

The upper and lower bounds of a $100(1-\alpha)\%$ confidence interval for the population standardized mean difference δ are given by

$$\hat{\delta}_C - Z_{\alpha/2}\sqrt{\text{Var}\left(\hat{\delta}_C\right)} \leq \delta \leq \hat{\delta}_C + Z_{\alpha/2}\sqrt{\text{Var}\left(\hat{\delta}_C\right)}, \tag{7.3}$$

where $Z_{\alpha/2}$ is the two-tailed critical value of the standard normal distribution at significance level α. The confidence interval that is based on the combined estimator is narrower than the confidence intervals that are based on the other estimators (that is, g_{U_+}, $\hat{\delta}_S$, or $\hat{\delta}_D$).

Example 7.1 *Using the combined procedure to obtain an estimate and a 95% confidence interval for the population standardized mean difference*

The data set used for this example is the same data set used for Examples 6.1 and 6.2 (cited in Bushman & Wang, 1996). It was taken from previous meta-analytic studies of intoxicated aggression (Bushman, 1993; Bushman & Cooper,

1990). One explanation of intoxicated aggression is based on alcohol-related expectancies. In laboratory experiments, this expectancy interpretation of intoxicated aggression can be tested by comparing the level of aggression for participants who expect an alcoholic beverage but receive a nonalcoholic beverage (that is, placebo) with the level of aggression for participants who expect and receive a nonalcoholic beverage (that is, control). The placebo versus control comparison gives the pure effects of alcohol-related expectancies on aggression uncontaminated by the pharmacological effects of alcohol. The data for 13 studies that included alcohol and placebo groups are given in Table 7.1. Sample standardized mean differences could be calculated for the first nine studies. The remaining four studies did not contain enough information to calculate sample standardized mean differences but did contain information about the direction of the results.

Table 7.1 *Studies testing the effects of alcohol-related expectancies on aggression (Bushman & Wang, 1996)*

Study	n_E	n_C	d	p	Direction
1	19	19	1.028	< .05	+
2	24	24	0.903	< .05	+
3	24	24	0.552	< .05	+
4	24	24	0.328	ns	+
5	12	12	0.193	ns	+
6	24	24	-0.036	ns	-
7	24	24	-0.332	ns	-
8	15	15	-0.540	ns	-
9	12	12	-0.660	ns	-
10	16	16		ns	+
11	12	12		ns	+
12	10	10		ns	+
13	10	10		ns	-

Note: n_E = number of participants in the placebo group. n_C = number of participants in the control group. d = sample standardized mean difference. *ns* = nonsignificant at the .05 level.

To obtain the combined estimator for the population standardized mean difference, you can use the SAS/IML modules VOTEA and VOTERUN (Chapter 6, Appendix 6.2), the SAS macros COMPEFF and WAVGMETA (Chapter 5, Appendices 5.3 and 5.6, respectively), along with the SAS code in Appendix 7.1. The results are printed in Output 7.1.

Output 7.1 *Using the combined procedure to obtain an estimate and a 95% confidence interval for the population standardized mean difference*

METHOD	STUDY	ESTIMATE	LOWER	UPPER
Combined	13	0.219	0.021	0.417
Effect-size (gu)	9	0.207	-.005	0.418
Vote-count (DIR)	4	0.306	-.253	0.864

In Output 7.1, STUDY is the number of studies that are included in the analysis, "Combined" is the estimate based on the combined procedure, "Effect-size (gu)" is the estimate that is based on effect-size procedures that use Hedges' g_U, and "Vote-count (DIR)" is the vote-count estimate that is based on the proportion of positive sample standardized mean differences. As you can see in Output 7.1, the 95% confidence interval for δ (Equation 7.3) based on all 13 studies is [0.021, 0.417], whereas the 95% confidence interval based on the nine studies with effect-size estimates is [−0.005, 0.418]. The confidence interval based on all 13 studies is about 6.4% narrower than the confidence interval based on the nine studies with effect-size estimates.

7.3 Using the Combined Procedure to Estimate the Population Correlation Coefficient

The combined procedure for correlation coefficients also uses the effect-size and vote-counting procedures that are described earlier in this book (see Chapters 5 and 6, respectively). Out of k independent studies, suppose that the first m_1 studies

($m_1 < k$) report enough information to compute the sample correlation coefficients, that the next m_2 studies ($m_2 < k$) only report the statistical significance of results, and that the last m_3 studies ($m_3 < k$) only report the direction of results, where $m_1 + m_2 + m_3 = k$. You can use effect-size procedures (Hedges and Olkin, 1985; page 234) to obtain a maximum likelihood estimate $\hat{\rho}_M$ of the population correlation coefficient ρ for the first m_1 studies. You can use vote-counting procedures (see Chapter 6) to obtain a maximum likelihood estimator $\hat{\rho}_S$ of ρ using the second m_2 studies. You can also use vote-counting procedures to obtain another maximum likelihood estimator $\hat{\rho}_D$ of ρ using the last m_3 studies. The respective variances for $\hat{\rho}_M$, $\hat{\rho}_S$, and $\hat{\rho}_D$ are $\mathrm{Var}(\hat{\rho}_M)$ (Hedges and Olkin, 1985, page 234), $\mathrm{Var}(\hat{\rho}_S)$ (Equation 6.8, where p_i is the proportion of significant positive correlations) and $\mathrm{Var}(\hat{\rho}_D)$ (Equation 6.8, where p_i is the proportion of positive correlations). The combined estimator of ρ is given by

$$\hat{\rho}_C = \frac{\hat{\rho}_M/\mathrm{Var}(\hat{\rho}_M)+\hat{\rho}_S/\mathrm{Var}(\hat{\rho}_S)+\hat{\rho}_D/\mathrm{Var}(\hat{\rho}_D)}{1/\mathrm{Var}(\hat{\rho}_M)+1/\mathrm{Var}(\hat{\rho}_S)+1/\mathrm{Var}(\hat{\rho}_D)}. \tag{7.4}$$

Because $\hat{\rho}_M$, $\hat{\rho}_S$, and $\hat{\rho}_D$ are consistent, the combined estimator $\hat{\rho}_C$ is consistent if (a) the studies that contain each type of information can be considered a random sample from the population of studies that contain the same type of information, and (b) m_1, m_2, and m_3 are all sufficiently large and all converge to infinity at the same rate.

The variance of the combined estimator is given by

$$\mathrm{Var}(\hat{\rho}_C) = \frac{1}{1/\mathrm{Var}(\hat{\rho}_M)+1/\mathrm{Var}(\hat{\rho}_S)+1/\mathrm{Var}(\hat{\rho}_D)}. \tag{7.5}$$

Because $\frac{1}{\mathrm{Var}(\hat{\rho}_M)} > 0$, $\frac{1}{\mathrm{Var}(\hat{\rho}_S)} > 0$, and $\frac{1}{\mathrm{Var}(\hat{\rho}_D)} > 0$, the combined estimator $\hat{\rho}_C$ is more efficient (that is, has smaller variance) than $\hat{\rho}_M$, $\hat{\rho}_S$, and $\hat{\rho}_D$.

The upper and lower bounds of a $100(1-\alpha)\%$ confidence interval for the population correlation coefficient ρ are given by

$$\hat{\rho}_C - Z_{\alpha/2}\sqrt{\mathrm{Var}(\hat{\rho}_C)} \le \rho \le \hat{\rho}_C + Z_{\alpha/2}\sqrt{\mathrm{Var}(\hat{\rho}_C)}. \tag{7.6}$$

The confidence interval that is based on the combined procedure is narrower than the confidence intervals that are based on other estimators (that is, $\hat{\rho}_M$, $\hat{\rho}_S$, and $\hat{\rho}_D$).

Example 7.2 *Using the combined procedure to obtain an estimate and a 95% confidence interval for the population correlation coefficient*

The data set used for this example is the same data set used for Examples 6.3 and 6.4 (see Gerrard, Gibbons, and Bushman, 1996). The studies were conducted to test the hypothesis that perceived risk to human immunodeficiency virus (HIV) infection motivates preventative sexual behaviors (for example, using condoms). Table 7.2 contains the results from 36 independent samples of participants. Pearson product-moment correlation coefficients were calculated for 12 of the 36 samples (that is, samples 1–12). The remaining 24 samples (that is, samples 13–36) did not contain enough information to calculate Pearson product-moment correlation coefficients but did contain information about the direction and/or significance of the results.

Table 7.2 *Studies correlating perceived vulnerability to HIV infection with precautionary sexual behaviors*

Study	n	r	p	Direction
1	294	.240	<.05	+
2	60	.240	ns	+
3	578	.123	<.05	+
4	459	.069	ns	+
5	241	-.050	ns	-
6	866	-.062	ns	-
7	493	-.130	<.05	-
8	212	-.162	<.05	-
9	272	-.210	<.05	-
10	147	-.260	<.05	-
11	481	-.342	<.05	-
12	99	-.346	<.05	-
13	1,127		<.05	+
14	312		<.05	+
15	99		ns	+
16	432		ns	+
17	637		ns	+
18	909		ns	+
19	580		ns	-
20	544		ns	-
21	501		ns	-
22	402		ns	-
23	266		ns	-
24	205		ns	-
25	197		ns	-
26	195		ns	-
27	152		ns	-
28	105		ns	-
29	111		ns	-
30	114		<.05	-
31	129		<.05	-
32	206		<.05	-
33	466		<.05	-
34	513		<.05	-
35	602		<.05	-
36	964		<.05	-

Note: ns = nonsignificant at the .05 level; + = positive result (that is, in predicted direction); - = negative result (that is, in opposite direction). r = Pearson product-moment correlation coefficient. n = total sample size.

To obtain the combined estimator for the population correlation coefficient, you need to use the SAS/IML modules VOTECDF1 and VOTEB (Appendices 6.1 and 6.3, respectively), the SAS macro MAXCORR(*nn,rr,level*) (Appendix 7.2), and the SAS code in Appendix 7.3. In the SAS macro MAXCORR, *nn* is the sample size for each study, *rr* is the sample correlation for each study, and *level* is the confidence level for the confidence interval based on the combined estimator. The results are printed in Output 7.2

Output 7.2 *Using the combined procedure to obtain an estimate and a 95% confidence interval for the population correlation coefficient*

METHOD	STUDY	ESTIMATE	LOWER	UPPER
Combined	36	-.041	-.061	-.021
MLE	12	-.066	-.096	-.036
Vote-count (SIG)	9	-.022	-.060	0.017
Vote-count (DIR)	15	-.023	-.058	0.012

In Output 7.2, STUDY is the number of studies included in the analysis, "Combined" is the estimate based on the combined procedure, "MLE" is the maximum likelihood estimate based on effect-size procedures, "Vote-count (SIG)" is the vote-count estimate based on the proportion of significant positive sample correlations, and "Vote-count (DIR)" is the vote-count estimate based on the proportion of positive sample correlations. As you can see in Output 7.2, the 95% confidence interval for ρ (Equation 6.6) based on all 36 studies is [-.061, -.021], whereas the 95% confidence interval that is based on the 12 studies with effect-size estimates is [-.096, -.036]. The confidence interval based on all 36 studies is about 33% narrower than the confidence interval that is based on the 12 studies with effect-size estimates.

7.4 Conclusions

The combined procedure that was described in this chapter has at least five advantages over conventional procedures for dealing with the problem of missing effect-size estimates (Bushman & Wang, 1996). First, the combined procedure uses all the information available from studies in a research synthesis; most other procedures ignore information about the direction and statistical significance of results. Second, the combined estimator is generally consistent. None of the conventional procedures for handling effect-size estimates are consistent; they all either overestimate or underestimate the population effect size. Third, the combined estimator is more efficient (that is, has small variance) than either the effect-size or vote-counting estimators. Fourth, the variance of the combined estimator is known. All of the conventional procedures for handling missing estimates either artificially inflate or artificially deflate the variance of effect-size estimates. Fifth, the combined procedure gives weight to all studies proportional to the Fisher information that they provide. In summary, the combined procedure is the method of choice for dealing with missing effect-size estimates when some studies do not provide enough information to compute effect-size estimates but do provide information about the direction and/or statistical significance of results. You can easily calculate a combined estimate and confidence interval for the population standardized difference and for the population correlation coefficient using the SAS programs in this book.

7.5 References

Buck, S. F. (1960). A method of estimation of missing values in multivariate data suitable for use with an electronic computer. *Journal of the Royal Statistical Society, Series B, 22*, 302–303.

Bushman, B. J. (1993). Human aggression while under the influence of alcohol and other drugs: An integrative research review. *Current Directions in Psychological Science, 2*, 148–152.

Bushman, B. J., & Cooper, H. M. (1990). Effects of alcohol on human aggression: An integrative research review. *Psychological Bulletin, 107*, 341–354.

Bushman, B. J., & Wang, M. C. (1995). A procedure for combining sample correlation coefficients and vote counts to obtain an estimate and a confidence interval for the population correlation coefficient. *Psychological Bulletin, 117*, 530–546.

Bushman, B. J., & Wang, M. C. (1996). A procedure for combining sample standardized mean differences and vote counts to estimate the population standardized mean difference in fixed effects models. *Psychological Methods, 1*, 66–80.

Draper, N. R., & Smith, H. (1966). Applied regression analysis. New York: Wiley.

Gerrard, M., Gibbons, F. X., & Bushman, B. J. (1996). Relation between perceived vulnerability to HIV and precautionary sexual behavior. *Psychological Bulletin, 119*, 390–409.

Hedges, L. V. (1982). Estimation of effect sizes from a series of independent experiments. *Psychological Bulletin, 92*, 490–499.

Hedges, L. V. & Olkin, I. (1985). *Statistical methods for meta-analysis.* New York: Academic Press.

Pigott, T. D. (1994). Methods for handling missing data in research synthesis. In H. Cooper & L. V. Hedges (Eds.), *Handbook of research synthesis* (pp. 163–175). New York: Russell Sage Foundation.

7.6 Appendices

Appendix 7.1: SAS Code for Example 7.1

```
options nodate nocenter pagesize=54 linesize=80 pageno=1;
libname ch7 "d:\metabook\ch7\dataset";
libname ch6 "d:\metabook\ch6\dataset";
/***********************************************************/
/*  Meta-analysis for studies that reported enough    */
/*   information to calculate effect-size estimates    */
/***********************************************************/
%covtefst(ch7.ex71a,temp);
%wavgmeta(temp,combine1,3,0.95);
data combine1;
   set combine1;
   method = "Effect-size (gu)";
   study = 9;
   variance = stdwavg * stdwavg;
   keep method study estimate variance lower upper;
run;
/***********************************************************/
/*  Meta-analysis for studies that only reported     */
/*   information about the direction of the results    */
/***********************************************************/

/* VOTERUN MACRO is defined in Appendix 6.4 */

%voterun(ch7.ex71b,"Mean",0.5,0.95);
data combine2;
   set voteaout;
   method = "Vote-count (DIR)";
   study = 4;
   estimate = delta;
   variance = vdelta;
   keep method study estimate variance lower upper;
data temp;
   set combine1 combine2;
   eff = estimate / variance;
   veff = 1 / variance;
   keep eff veff;
proc means data = temp noprint;
   var eff veff;
   output out = tempout sum = s1 s2;
```

```
/***********************************************************/
/* The combined procedure                                  */
/***********************************************************/
data combine3;
   set tempout;
   method = "Combined          ";
   study = 13;
   estimate = s1 / s2;
   variance = 1 / s2;
   lower = estimate + probit(0.025) * sqrt(variance);
   upper = estimate + probit(0.975) * sqrt(variance);
   keep method study estimate variance lower upper;
data result; set combine3 combine1 combine2;
proc print data = result noobs;
   var method study estimate lower upper;
   title;
   format estimate 5.3 lower upper 5.3;
run;
```

Appendix 7.2: SAS Macro for Obtaining the Pearson Product-Moment Correlation Coefficient Based on the Method of Maximum Likelihood

```
options nodate nocenter pagesize=54 linesize=80 pageno=1;
libname aa "d:\metabook\ch6\dataset";
proc iml;
   start maxcorr(nn,rr,level);
/*****************************************************/
/*  Input Variables:                              */
/*     NN: Study sample size                      */
/*     RR: Sample correlation coefficient         */
/*     LEVEL: Confidence level of the confidence interval*/
/*                                                */
/*  Output Variables:                             */
/*     ESTIMATE: Maximum likelihood effect-size estimate */
/*     VARIANCE: Maximum likelihood variance estimate    */
/*     LOWER: Lower bound for the confidence interval    */
/*     UPPER: Upper bound for the confidence interval    */
/*     LEVEL: Confidence level of the confidence interval*/
/*****************************************************/
/*****************************************************/
/*     Compute the initial value for Newton iteration */
/*****************************************************/
   xold = 0.5*(max(rr)+min(rr));
/*****************************************************/
/*     Newton iteration to compute the maximum       */
/*     likelihood effect-size estimate               */
/*****************************************************/
   abserr=1;
   k=1;
   do while ((abserr > 1d-5) & (k < 30));
      fold=0;
      fpold=0;
      do I = 1 to nrow(nn);
         fold=fold+(nn[I,1]*(rr[I,1]-xold))/(1-
```

```
rr[I,1]*xold);
            fpold=fpold+nn[I,1]*(rr[I,1]*rr[I,1]-1)/
                ((1-rr[I,1]*xold)*(1-rr[I,1]*xold));
        end;
        xnew = xold - fold / fpold;
        abserr = abs(xnew-xold);
        k = k + 1;
        xold = xnew;
    end;
    estimate = xnew;
    variance = (1-xnew*xnew)*(1-xnew*xnew)/sum(nn);
    lower = xnew + probit(0.5*level)*sqrt(variance);
    upper = xnew + probit(1-0.5*level)*sqrt(variance);
    create mcorrout var {estimate variance lower upper};
    append;
    close mcorrout;
    finish maxcorr;
    reset storage=aa.imlrout;
    store module=(maxcorr);
run;
quit;
```

Appendix 7.3: SAS Code for Example 7.2

```
options nodate nocenter pagesize=54 linesize=80 pageno=1;
libname aa "d:\metabook\ch6\dataset";
libname ch7 "d:\metabook\ch7\dataset";
/*******************************************************/
/*  Meta-analysis for studies that reported enough     */
/*  information to calculate Pearson product-moment    */
/*  correlations                                       */
/*******************************************************/
proc iml;
   reset nolog;
   reset storage=aa.imlrout;
   load module=(maxcorr);
   use ch7.data72a;
   read all;
   level=0.05;
   run maxcorr(nn,rr,level);
run;
data combine1;
   method="MLE             ";
   study = 12;
   set mcorrout;
/*********************************************************/
/*  Meta-analysis for studies that only reported the    */
/*  significance and direction of sample correlations   */
/*********************************************************/
/* VOTERUN MACRO is defined in Appendix 6.4 */
%voterun(ch7.data72b,"corr",0.05,0.95);
data combine2;
   method="Vote-count (SIG)";
   study=9;
   set votebout;
   estimate = rho;
   variance = vrho;
   keep method study estimate variance lower upper;
/*********************************************************/
/*  Meta-analysis for studies that only reported the    */
/*  direction of sample correlations                    */
/*********************************************************/
%voterun(ch7.data72c,"corr",0.50,0.95);
data combine3;
   method="Vote-count (DIR)";
   study=15;
   set votebout;
   estimate = rho;
   variance = vrho;
   keep method study estimate variance lower upper;
   set votebout;
```

```
/*********************************************************/
/* The combined procedure                               */
/*********************************************************/
data temp;
    set combine1 combine2 combine3;
    eff = estimate / variance;
    veff = 1 / variance;
    keep eff veff;
proc means data = temp noprint;
    var eff veff;
    output out = tempout sum = s1 s2;
data combine4;
    set tempout;
    method="Combined        ";
    study=36;
    estimate = s1 / s2;
    variance = 1 / s2;
    lower = estimate + probit(0.025) * sqrt(variance);
    upper = estimate + probit(0.975) * sqrt(variance);
    keep method study estimate variance lower upper;
data result; set combine4 combine1 combine2 combine3;
proc print data = result noobs;
    var method study estimate lower upper;
    title;
    format estimate 5.3 lower upper 5.3;
run;
```

chapter 8

Fixed-Effects Models in Meta-Analysis

8.1 Introduction

As was noted in previous chapters, you should not combine effect-size estimates unless they are homogeneous or similar in magnitude. If the variation in effect-size estimates is greater than you would expect by random chance, you should not combine the effect-size estimates at all. You can formally test whether effect-size estimates are too heterogeneous to combine. A statistically significant heterogeneity test implies that variation in effect-size estimates between studies is significantly larger than you would expect by random chance.

If the heterogeneity test is significant, the "extra" variation in effects between-studies might be due to systematic differences between studies. You can enter these study characteristics as variables into an analysis of variance (ANOVA) or regression model, depending on whether the study characteristics are categorical or continuous, to see if they account for the excess variation in effect-size estimates. You can also analyze whether study characteristics moderate effect-size estimates even when the heterogeneity test is nonsignificant, but such analyses should be based on *a priori* (planned) hypotheses (that is, you should not go "fishing" for significant moderator effects).

Study characteristics can be treated as fixed or random variables. The fixed-effects model assumes that the population effect size is a single fixed value, whereas the random-effects model assumes that the population effect size is a randomly distributed variable with its own mean and variance. You should use a fixed-effects model if the variation in effect-size estimates between studies is due to a few simple study characteristics. When you use a fixed-effects model, you can make generalizations to a universe of studies with similar study characteristics. You should use a random-effects model if the differences between studies are too complicated to be captured by a few study characteristics. When you use a random-effects model, you can make generalizations to a universe of such diverse studies. Although generalizability is higher for random-effects models than for fixed-effects models, statistical power is higher for fixed-effects models than for random-effects models, (Rosenthal, 1995). Consequently, confidence intervals for effect sizes are narrower for fixed-effects models than for random-effects models. This chapter describes fixed-effects models in meta-analysis, whereas Chapter 9 describes random-effects models in meta-analysis:

Section 8.2   discusses the distinction between fixed- and random-effects models in individual experiments to help clarify the distinction between fixed- and random-effects models in meta-analysis.

Section 8.3   discusses the distinction between fixed- and random-effects models in meta-analysis.

Section 8.4   focuses on testing the moderating effects of categorical study characteristics in fixed-effects ANOVA models; both one-factor and two-factor ANOVA models are considered.

Section 8.5   focuses on testing the moderating effects of continuous study characteristics in fixed-effects regression models.

Section 8.6   discusses a method for quantifying the amount of variation explained by study characteristics.

Section 8.7   offers some conclusions.

Before proceeding with our discussion, a word of caution is in order. Like all tests, the statistical significance of the homogeneity test depends on the number of studies that are included in the meta-analysis. If the number of studies is very small, the homogeneity test might be nonsignificant simply because the test had low statistical power. If the number of studies is very large, the homogeneity test might be significant even if the effects are quite homogeneous. Thus, theoretical factors should play a substantial role in the evaluation of whether to use a fixed-effects or random-effects model. We strongly suggest that you use of both types of models. You can then use sensitivity analysis to determine whether the conclusions differ as a function of the type of model used.

8.2 Fixed- and Random-Effects Models in Individual Experiments

In an individual experiment, you can select the levels of a factor deliberately or randomly. The way in which you select the factor levels has implications for the inferences that can be made. If you select the factor levels deliberately, the inference made is restricted to the factor levels used in the experiment. If you select the factor levels randomly, the inference made is not restricted to the factor levels used in the experiment.

If you select the factor levels deliberately, the mathematical model of the experiment is called a fixed-effects model. It is called a fixed-effects model because the treatments remain fixed or constant from one replication of the experiment to another. If you select the factor levels randomly, the mathematical model of the experiment is called a random-effects model. It is called a random-effects model because the treatments tested in one replication of the experiment may differ from those tested in another replication.

The fixed-effects model for a single-factor experiment is

$$Y_{ij} = \mu + \alpha_i + \varepsilon_{ij}, \tag{8.1}$$

where Y_{ij} is the jth observation $(j = 1, \dots, n)$ in the ith treatment group $(i = 1, \dots, a)$; μ is a constant, the mean of all possible experiments using the a deliberately selected treatments; α_i is a constant for the ith treatment group, the deviation from the population mean due to the ith treatment, $\sum_{i=1}^{a} \alpha_i = 0$; and ε_{ij} is the random effect that contains all uncontrolled sources variability (Dowdy & Weardon, 1983, p. 288). The ε_{ij}'s are independently and normally distributed with mean zero and variance σ_i^2. That is, the ε_{ij}'s are independent and have distribution $N(0, \sigma_i^2)$.

The random-effects model for a single-factor experiment is

$$Y_{ij} = \mu + A_i + \varepsilon_{ij}, \tag{8.2}$$

where Y_{ij} is the jth observation $(j = 1, \dots, n)$ in the ith treatment group $(i = 1, \dots, a)$; μ is a constant, the mean of all experiments involving all possible treatments of the type being tested; A_i is a constant for the ith treatment group, a random deviation from the population mean; and ε_{ij} is the random effect that contains all uncontrolled sources variability (Dowdy & Weardon, 1983, p. 288). The A_i's are independent and have distribution $N(0, \sigma_A^2)$, and the ε_{ij}'s are independent and have distribution $N(0, \sigma_i^2)$. In both fixed- and random-effects models, it is assumed that the experimental units (for example, participants) are selected at random and are assigned to treatments at random (Dowdy & Weardon, 1983, p. 289).

In a random-effects model, the researcher usually is not interested in testing hypotheses, constructing confidence intervals, or contrasting means. Instead, the researcher is interested in estimating components of variance; he or she wants to know how much of the variance in the experiment is due to true differences in treatment means, and how much is due to random error about treatment means (that is, sampling error).

The hypotheses tests differ for fixed- and random-effects models. In a single factor experiment, the null hypothesis for a fixed-effects model is

$$H_0 : \alpha_1 = \ldots = \alpha_a, \qquad (8.3)$$

whereas the null hypothesis for a random-effects model is

$$H_0 : \sigma_A^2 = 0 \qquad (8.4)$$

(Dowdy & Weardon, 1983, p. 288). To help the researcher estimate the components of variance in an experiment, the ANOVA table contains an expected mean square column. In the fixed-effects model, the mean square between factors ($MS_{BETWEEN}$) estimates $\sigma^2 + n\sum_{i=1}^{a} \alpha_i^2 / (a-1)$, whereas in the random-effects model, $MS_{BETWEEN}$ estimates $\sigma^2 + n\sigma_A^2$ (Dowdy & Weardon, 1983, p. 289). In both fixed- and random-effects models, the error mean square (MS_{ERROR}) estimates σ^2.

8.3 Fixed- and Random-Effects Models in Meta-Analysis

Hedges and Olkin (1985) developed fixed- and random-effects models for effect sizes in meta-analysis, analogous to those for primary data in individual experiments. In an individual experiment, use a fixed-effects model if the variation between experimental units (for example, participants) is considered to be the result of sampling error alone; use a random-effects model if the variation between experimental units is considered to be the result of both sampling error and true differences in treatment means. In a meta-analysis, use a fixed-effects model if the variation in effect sizes between studies is considered to be the result of sampling error alone; use a random-effects model if the variation in effect sizes between studies is considered to be the result of both sampling error and variability in the population effect size.

This chapter focuses on fixed-effects models in meta-analysis. When between-studies effect-size variation is treated as fixed, the only source of variation treated as random is the within-studies sampling variation. A statistically significant heterogeneity test implies that the variation in effect-size estimates between studies is significantly larger than you would expect by chance alone. Hedges (1994) recommended that meta-analysts should try to explain this "extra" variation in effect-size estimates between studies by entering known study characteristics in an ANOVA or regression model.

The next section describes how studies in a meta-analysis might differ. These differences between studies, called collectively "study characteristics," can be considered either categorical or continuous variables. Section 8.4 focuses on testing the moderating effects of categorical study characteristics in fixed-effects ANOVA models, whereas Section 8.5 focuses on testing the moderating effects of continuous study characteristics in fixed-effects regression models.

8.4 Testing the Moderating Effects of Categorical Study Characteristics in ANOVA Models

Although all studies in a meta-analysis investigate the same conceptual variables, in the individual studies researchers often use different operational definitions of the conceptual variables (see Chapter 1, Section 1.4). Studies may also differ on other characteristics, such as source characteristics (for example, publication year; publication status; sex of first author), participant characteristics (for example, population sampled; sex, age, race of participants), and design characteristics (for example, between- or within-subjects design; prospective, cross-sectional, or retrospective design; use of random assignment, use of double-blind procedure). We use the term "study characteristics" to refer to all of the different ways in which studies in a meta-analysis might differ. This section focuses on testing the moderating effects of categorical study characteristics in fixed-effects ANOVA models. We consider ANOVA models with one and two categorical factors.

8.4.1 Fixed-Effects ANOVA Models with One Categorical Factor

Suppose that you are interested in testing the moderating effect of one categorical study characteristic with p independent groups, with m_1 effects in group 1, m_2 effects in group 2, ..., and m_p effects in group p. Let θ_{ij} denote the jth effect-size parameter in the ith group, and let $\hat{\theta}_{ij}$ denote its estimate with sampling variance $\mathrm{Var}(\hat{\theta}_{ij})$. Table 8.1 gives the notation for a fixed-effects model with one categorical factor (Hedges, 1994, p. 287).

Table 8.1 *Effect-size estimates and sampling variances for p groups of studies.*

	Effect-size estimates	Variance estimates
Group 1		
Study 1	$\hat{\theta}_{11}$	$\text{Var}(\hat{\theta}_{11})$
Study 2	$\hat{\theta}_{12}$	$\text{Var}(\hat{\theta}_{12})$
$\vdots$	$\vdots$	$\vdots$
Study m_1	$\hat{\theta}_{1m_1}$	$\text{Var}(\hat{\theta}_{1m_1})$
Group 2		
Study 1	$\hat{\theta}_{21}$	$\text{Var}(\hat{\theta}_{21})$
Study 2	$\hat{\theta}_{22}$	$\text{Var}(\hat{\theta}_{22})$
$\vdots$	$\vdots$	$\vdots$
Study m_2	$\hat{\theta}_{2m_2}$	$\text{Var}(\hat{\theta}_{2m_2})$
$\vdots$	$\vdots$	$\vdots$
Group p		
Study 1	$\hat{\theta}_{p1}$	$\text{Var}(\hat{\theta}_{p1})$
Study 2	$\hat{\theta}_{p2}$	$\text{Var}(\hat{\theta}_{p2})$
$\vdots$	$\vdots$	$\vdots$
Study m_p	$\hat{\theta}_{pm_p}$	$\text{Var}(\hat{\theta}_{pm_p})$

8.4.1.1 Tests of Heterogeneity

In the analysis of variance for a one-factor experiment, the total sum of squares can be partitioned into two components: (a) sum of squares between groups of participants and (b) sum of squares within groups of participants. In symbols,

$$SS_{TOTAL} = SS_{BETWEEN} + SS_{WITHIN}. \tag{8.5}$$

Similarly, in the analysis of variance for a meta-analysis, the total heterogeneity statistic can be partitioned into two components: (a) heterogeneity between groups of studies and (b) heterogeneity within groups of studies. In symbols,

$$Q_{TOTAL} = Q_{BETWEEN} + Q_{WITHIN}. \tag{8.6}$$

To test the null hypothesis that there is no variation in group mean effect sizes,

$$H_0: \theta_{1+} = \theta_{2+} = \ldots = \theta_{p+}, \tag{8.7}$$

you use the statistic

$$Q_{BETWEEN} = \sum_{i=1}^{p} w_{i+} \left(\hat{\theta}_{i+} - \hat{\theta}_{++} \right)^2, \tag{8.8}$$

where $w_{i+} = \sum_{j=1}^{m_i} w_{ij}$, the weight $w_{ij} = 1/\mathrm{Var}\left(\hat{\theta}_{ij}\right)$,

$$\hat{\theta}_{i+} = \frac{\sum_{j=1}^{m_i} w_{ij} \hat{\theta}_{ij}}{\sum_{j=1}^{m_i} w_{ij}}, \tag{8.9}$$

is the weighted mean of the effect-size estimates in the ith group, and

$$\hat{\theta}_{++} = \frac{\sum_{i=1}^{p} \sum_{j=1}^{m_i} w_{ij} \hat{\theta}_{ij}}{\sum_{i=1}^{p} \sum_{j=1}^{m_i} w_{ij}} = \frac{\sum_{i=1}^{p} w_{i+} \hat{\theta}_{i+}}{\sum_{i=1}^{p} w_{i+}}, \tag{8.10}$$

is the grand weighted mean (Hedges, 1994, p. 289). You reject the null hypothesis in Equation 8.3 at significance level α if $Q_{BETWEEN}$ exceeds the $100(1-\alpha)\%$ critical value of the chi-square distribution with $p-1$ degrees of freedom. The statistic $Q_{BETWEEN}$ yields a direct omnibus test for between-group differences in mean effects. It is analogous to the omnibus F test for between-group differences in a one-factor ANOVA for an individual experiment.

To test the null hypothesis that there is no variation in population effects within the groups of studies,

$$H_0: \begin{matrix} \theta_{11} = \ldots = \theta_{1m_1} = \theta_{1+} \\ \vdots \qquad \vdots \qquad \vdots \\ \theta_{p1} = \ldots = \theta_{pm_p} = \theta_{p+} \end{matrix}, \tag{8.11}$$

you use the statistic

$$Q_{WITHIN} = \sum_{i=1}^{k} \sum_{j=1}^{m_i} w_{ij} \left(\hat{\theta}_{ij} - \hat{\theta}_{i+} \right)^2 \tag{8.12}$$

(Hedges, 1994, p. 290). You reject the null hypothesis in Equation 8.11 at significance level α if Q_{WITHIN} exceeds the $100(1-\alpha)\%$ critical value of the chi-square distribution with $k-p$ degrees of freedom.

Hedges (1994) has pointed out that although Q_{WITHIN} provides an overall test of within-group heterogeneity of effects, it is actually the sum of p separate within-group heterogeneity statistics. That is, $Q_{WITHIN} = Q_{WITHIN_1} + Q_{WITHIN_2} + \ldots + Q_{WITHIN_p}$. The individual heterogeneity tests are useful in determining which groups are the major sources of within-group heterogeneity and which groups have relatively homogeneous effects.

Table 8.2 summarizes the various sources of heterogeneity in a meta-analysis with one categorical study characteristic, as well as the test statistics and corresponding degrees of freedom for each source of heterogeneity (Hedges, 1994, p. 290).

Table 8.2 *Heterogeneity summary table for a fixed-effects model with one categorical factor*

Source	Q Statistic	df
Between groups	$Q_{BETWEEN}$	$p-1$
Within groups	Q_{WITHIN}	$k-p$
Within group 1	Q_{WITHIN_1}	$m_1 - 1$
Within group 2	Q_{WITHIN_2}	$m_2 - 1$
$\vdots$	$\vdots$	$\vdots$
Within group p	Q_{WITHIN_p}	$m_p - 1$
Total	Q_{TOTAL}	$k-1$

Note: $k = m_1 + m_2 + \ldots + m_p$. $df =$ degrees of freedom.

Example 8.1 *Heterogeneity tests for a one-factor model*

The data for this example were obtained from a meta-analysis by Bettencourt and Miller (1996) on sex differences in aggression as a function of provocation. Participants in these studies were provoked by a confederate or experimenter who verbally or physically attacked them. The verbal attacks consisted of insults and the physical attacks consisted of noxious stimuli such as electric shocks and noise blasts. The studies in this meta-analysis used three different types of aggression measures: physical, money loss, and evaluation. The most common laboratory paradigm for measuring physical aggression is the "aggression machine" (Buss, 1961). In this paradigm, participants give noxious stimuli (for example, electric shocks, noise blasts) to a confederate whenever he or she makes an error on a learning task. The intensity and/or duration of the noxious stimuli that participants give the confederate is used to measure physical aggression. In one variation of this paradigm, participants subtract money from the confederate whenever he or she makes an error on a learning task.

In other studies, participants evaluate a confederate or experimenter. In one study (Rohsenow & Bachorowski, 1984), for example, each participant was told to trace a circle as slowly as possible. After this task was completed, the experimenter burst in the room, introduced himself as the supervisor who had been observing through a one-way mirror, and contemptuously stated, "Obviously, you don't follow instructions. You were supposed to trace the circle as slowly as possible without stopping but you clearly didn't do this. Now I don't know if we can use your data." The experimenter paused, then continued (interrupting the participant if he or she tried to respond), "Do it over again." After the experiment, the participant completed an evaluation form for each member of the lab staff, including the obnoxious experimenter. The evaluations were placed in a sealed envelope and were allegedly sent to the principal investigator to be used in future hiring decisions. The participant could, therefore, harm the experimenter's chances of being rehired by evaluating him in a negative manner.

The data from the Bettencourt and Miller (1996) meta-analysis are depicted in Table 8.3. We will use this data set for one- and two-factor ANOVA models.

Table 8.3 Meta-analysis on sex differences in aggression as a function of provocation.

Study	Aggression	n_P	Provocation g	Provocation Var(g)	n_C	Control g	Control Var(g)
1	1	40	0.98	0.11	40	0.78	0.11
2	2	10	0.87	0.44	10	0.15	0.40
3	1	100	0.78	0.04	100	-0.58	0.04
4	1	20	0.64	0.21	20	0.37	0.20
5	2	120	0.53	0.03	60	0.08	0.07
6	1	40	0.52	0.10	40	0.63	0.11
7	1	40	0.46	0.10	40	0.52	0.10
8	1	240	0.42	0.19	240	-0.16	0.03
9	2	96	0.38	0.04	96	0.48	0.04
10	2	25	0.30	0.16	25	-0.15	0.16
11	1	194	0.28	0.02	194	0.49	0.02
12	1	12	0.21	0.33	12	0.15	0.34
13	1	40	0.19	0.10	40	0.07	0.10
14	1	10	0.00	0.40	10	0.00	0.40
15	2	80	0.00	0.05	80	0.00	0.05
16	2	34	0.00	0.12	34	0.00	0.12
17	2	40	0.00	0.10	40	0.00	0.10
18	2	40	0.00	0.10	40	0.00	0.10
19	2	39	-0.02	0.10	39	2.03	0.15
20	1	12	-0.02	0.33	12	0.08	0.34
21	1	12	-0.03	0.33	12	0.02	0.34
22	1	20	-0.09	0.20	20	1.13	0.23
23	1	12	-0.10	0.33	12	0.60	0.35
24	1	10	-0.11	0.40	10	2.31	0.67
25	3	60	-0.29	0.07	60	-0.02	0.07
26	2	72	-0.44	0.08	72	0.39	0.04
27	1	28	-0.67	0.20	27	0.40	0.15
28	3	40	-0.69	0.10	40	-0.10	0.10
29	2	20	-0.84	0.22	20	0.71	0.21
30	2	8	-1.19	0.59	8	0.61	0.52
31	1	10	-1.20	0.47	10	1.87	0.58
32	2	14	-2.52	1.65	14	0.31	0.29

Note: Aggression: 1 = physical, 2 = evaluation, 3 = money loss; g = Hedges' estimate of the population standardized mean difference, positive values of g indicate that males were more aggressive than females; Var(g) = variance of g; n_P = number of participants in provocation condition; n_C = number of participants in control condition.

The SAS macro WITHIN(*indata,outdata,eff,veff,ftname,nlevels*) in Appendix 8.1 was used to calculate the heterogeneity summary statistics for the ANOVA model that included type of aggression measure (AGGR) as the only factor. *Indata* and *outdata* are the names of the input and output data sets, respectively. You can access the input data set for this example, CH8.EX81, via SAS Online Samples (see "Using This Book"). In the SAS data set CH8.EX81, the aggression measure variable is called AGGR and the provocation variable is called PROV. The effect-size estimate is *eff*, and *veff* is its corresponding estimated variance. If the estimated effect-size variances are not reported, but the sample sizes are reported, you can obtain the variances using the SAS macros in Chapter 5, Appendix 5.3. The name of the factor in the ANOVA model is *ftname*, and *nlevels* is the number of levels of the factor. In this example, *ftname* is AGGR (that is, type of aggression measure used), and *nlevels* equals 3 (that is, physical, evaluation, money loss).

The statement `%let varlist1 = aggr` assigns the macro variable VARLIST1 the value AGGR (see Chapter 2, section 2.6). The macro variable VARLIST1 is used for the CLASS statement in the GLM procedure. The statement `%let varlist1 = aggr` assigns the macro variable VARLIST2 the value AGGR. The macro variable VARLIST2 is used for the MODEL statement in the GLM procedure.

The results given in Output 8.1 were produced by the following SAS code:

```
options nodate nocenter pagesize=54 linesize=80 pageno=1;
libname ch8 "d:\metabook\ch8\dataset";
%let varlist1 = aggr;
%let varlist2 = aggr;
%within(ch8.ex81,ch8.ex81outa,eff,veff,aggr,3);
proc format;
    value aa  1 = "Between Groups         "
              2 = "Within Groups          "
              3 = "  Within (Physical)    "
              4 = "  Within (Evaluation)  "
              5 = "  Within (Money Loss)  "
              6 = "Corrected Total        ";
proc print data=ch8.ex81outa noobs;
    title;
    format source aa. qstat pvalue 7.3;
run;
```

Output 8.1 ANOVA *table for heterogeneity summary statistics for the meta-analysis on sex differences in aggression*

SOURCE	DF	QSTAT	PVALUE
Between Groups	2	12.296	0.002
Within Groups	61	128.390	0.000
Within (Physical)	33	70.042	0.000
Within (Evaluation)	25	55.408	0.000
Within (Money Loss)	3	2.939	0.401
Corrected Total	63	140.685	0.000

As you can see in Output 8.1, the between-groups heterogeneity test is significant, indicating that the magnitude of effect-size estimates is not the same for the different measures of aggression. Section 8.4.1.2 describes how to conduct *a priori* (planned) and *post hoc* (unplanned) comparisons of group effects. The within-groups heterogeneity test is also significant, due primarily to heterogeneous effects for the physical and evaluation measure of aggression. Other factors might moderate sex differences in these two aggression measures.

The SAS macro WAVGEFF(*indata,outdata,level*) in Appendix 8.2 can compute the effect-size estimates and confidence intervals for fixed-effects ANOVA models with one categorical factor. *Indata* and *outdata* are the names of the input and output data sets, respectively. *Level* is the level of confidence for the confidence intervals (for example, use .95 for 95% confidence intervals). One artificial factor level is used to obtain the overall effect-size estimate. In this example, aggr = 4 was used to obtain the overall effect-size estimate for all three types of aggression measures combined. The results from the following SAS code are displayed in Output 8.2:

```
options nodate nocenter pagesize=54 linesize=80 pageno=1;
libname ch8 "d:\metabook\ch8\dataset";
data temp;
    set ch8.ex81;
    output;
    aggr = 4;
    output;
%let varlist = aggr;
%wavgeff(temp,ch8.ex81outb,.95);
proc format;
```

```
    value aa 1 = "Physical  "
             2 = "Evaluation"
             3 = "Money Loss"
             4 = "Over All  ";
proc print data=ch8.ex81outb noobs;
    var aggr estimate level lower upper;
    format aggr aa. estimate level lower upper 6.3;
run;
```

Output 8.2 *Sex difference 95% confidence intervals for different measures of aggression and for all measures combined*

AGGR	ESTIMATE	LEVEL	LOWER	UPPER
Physical	0.285	0.950	0.177	0.392
Evaluation	0.210	0.950	0.094	0.327
Money loss	-0.254	0.950	-0.535	0.027
Overall	0.214	0.950	0.138	0.290

Because the "Between groups" and "Within groups" heterogeneity tests were both significant (see Output 8.1), you should not use the overall measure of aggression in Output 8.2. For physical and evaluation aggression measures, males were more aggressive than females. For money-loss aggression measures, the amount of aggression that was displayed by males and females did not differ. The latter result must be interpreted with caution, however, because only two studies used money loss as a measure of aggression.

8.4.1.2 Comparisons among Group Mean Effects

If the independent variable in a single-factor experiment contains more than two levels, the omnibus F test indicates only that the group means are not all equal. *A priori* (planned) or *post hoc* (unplanned) comparisons are conducted to determine how the group means differ. Similarly, if the study characteristic variable in a meta-analysis contains more than two levels, the omnibus $Q_{BETWEEN}$ statistic indicates that the weighted group mean effect sizes are not all equal. *A priori* or *post hoc* comparisons are conducted to determine how the group mean effects differ.

8.4.1.2.1 *A priori contrasts among group mean effects.* To make *a prior* contrasts among group mean effects, you create a linear combination of the population mean effects:

$$\gamma = c_1\theta_{1+} + \ldots + c_p\theta_{p+},$$ (8.13)

where $c_1, \ldots, c_p$ are the contrast coefficients that satisfy the constraint $c_1 + \ldots + c_p = 0$ (Hedges, 1994, p. 292). The contrast coefficients are chosen to reflect a comparison of interest. For example, to compare mean effect for group 1 with the mean effect for group 2, use the contrast coefficients $c_1 = 1, c_2 = -1, c_3 = c_4 = \ldots = c_p = 0$. To compare the mean effects for groups 1 and 2 with the mean effect for group 3, use the contrast coefficients $c_1 = 0.5, c_2 = 0.5, c_3 = -1, c_4 = \ldots = c_p = 0$.

The contrast parameter γ in Equation 8.13 is estimated by a linear combination of sample effect means:

$$\hat{\gamma} = c_1\hat{\theta}_{1+} + \ldots + c_p\hat{\theta}_{p+}.$$ (8.14)

The estimated contrast $\hat{\gamma}$ has a normal sampling distribution with estimated variance

$$\mathrm{Var}(\hat{\gamma}) = c_1^2\mathrm{Var}(\hat{\theta}_{1+}) + \ldots + c_p^2\mathrm{Var}(\hat{\theta}_{p+})$$ (8.15)

(Hedges, 1994, p. 292). The $100(1-\alpha)\%$ confidence interval for the contrast parameter γ is given by

$$\hat{\gamma} - Z_{\alpha/2}\sqrt{\mathrm{Var}(\hat{\gamma})} \leq \gamma \leq \hat{\gamma} + Z_{\alpha/2}\sqrt{\mathrm{Var}(\hat{\gamma})}.$$ (8.16)

8.4.1.2.2 Post hoc contrasts among group mean effects. If several tests among group means are made at the same significance level α, then some of the tests will be significant by chance alone. Simultaneous test procedures have been developed to control the overall Type I error rate that might occur when several tests are performed. The Bonferroni procedure controls the overall Type I error rate by conducting each test at significance level α/c, where c is the number of tests conducted. Thus, the Bonferroni simultaneous confidence intervals are given by

$$\hat{\gamma}_i - Z_{\alpha/2c}\sqrt{\mathrm{Var}(\hat{\gamma}_i)} \le \gamma_i \le \hat{\gamma}_i + Z_{\alpha/2c}\sqrt{\mathrm{Var}(\hat{\gamma}_i)}, \tag{8.17}$$

for $i = 1,2,\ldots,c$.

Hedges and Olkin (1985, p. 161) proposed an alternative procedure that is based on the Scheffe multiple comparison procedure. With the Scheffe procedure, you reject the null hypothesis that the contrast γ (Equation 8.9) equals zero if the test statistic

$$X^2 = \hat{\gamma}^2 / \mathrm{Var}(\hat{\gamma}) \tag{8.18}$$

exceeds the $100(1-\alpha)\%$ critical value of the chi-square distribution with c' degree of freedom, where c' is the smaller of c (the number of contrasts) or $p-1$ (the number of groups minus one). The simultaneous confidence intervals for Scheffe's procedure are

$$\hat{\gamma}_i - Z_{\alpha/2c'}\sqrt{\mathrm{Var}(\hat{\gamma}_i)} \le \gamma_i \le \hat{\gamma}_i + Z_{\alpha/2c'}\sqrt{\mathrm{Var}(\hat{\gamma}_i)}, \tag{8.19}$$

for $i = 1,2,\ldots,c'$.

Example 8.2 *A priori and post hoc comparisons among group mean effects*

Recall from Example 8.1 that the between-group homogeneity statistic was significant, indicating that sex-differences were not the same for all measures of aggression. You can use the SAS macro FIXCONST(*indata,outdata,coeff,kind,level*) in

Appendix 8.3 to investigate which aggression measures differed from which other measures. *Indata* is the name of the output data file from the SAS macro WAVGEFF (see Appendix 8.2). It contains the effect-size estimates and corresponding variances that are needed to compute the comparisons among group mean effects. *Outdata* is the name of the output data file that contains the contrast estimates and corresponding upper and lower confidence interval bounds. *Coeff* is the file name of the data set that contains the coefficients for planned contrasts. *Coeff* is an empty data set if you want to do all possible pairwise comparisons. *Kind* is a categorical variable that specifies the type of contrast to be performed: Bonferroni (*kind*=1), Scheffe (*kind*=2), or *a priori* (*kind*=3). LEVEL is the level of confidence for the confidence interval (for example, use .95 for a 95% confidence interval).

The statement %let varlist = estimate variance in line 7 of the SAS code assigns the macro variable VARLIST the values ESTIMATE and VARIANCE. ESTIMATE and VARIANCE are the respective names for the effect-size estimate and its corresponding variance obtained from the SAS macro WAVGEFF in Appendix 8.2.

The following SAS code was used to obtain the Bonferroni contrasts printed in Output 8.3. Note that *kind*=1 in this example.

```
options nodate nocenter pagesize=54 linesize=80 pageno=1;
libname ch8 "d:\metabook\ch8\dataset";
data coeff;
data tmpin;
   set ch8.ex81outb;
   if (aggr ^= 4);
%let varlist = estimate variance;
%fixconst(tmpin,tmpout,coeff,1,0.95)
proc format;
   value aa 1 = "Physical vs. Evaluation   "
            2 = "Physical vs. Money loss   "
            3 = "Evaluation vs. Money loss";
proc print data=tmpout noobs;
   title;
   format type aa. contrast level lower upper 7.3;
quit;
```

Output 8.3 *Post hoc comparisons among aggression measures using Bonferroni's procedure*

TYPE	CONTRAST	LEVEL	LOWER	UPPER
Physical vs. Evaluation	0.074	0.950	-0.119	0.268
Physical vs. Money Loss	0.538	0.950	0.171	0.906
Evaluation vs. Money Loss	0.464	0.950	0.092	0.836

As you can see in Output 8.3, the confidence intervals for the physical versus money loss and evaluation versus money loss comparisons do not contain the value zero. The confidence interval for the physical versus evaluation comparison, however, does contain the value zero.

To make *post hoc* comparisons among group means with the Scheffe procedure, you use *kind*=2 rather than *kind*=1 in the SAS code. The results are printed in Output 8.4.

Output 8.4 *Post hoc comparisons among aggression measures using Scheffe's procedure*

TYPE	CONTRAST	LEVEL	LOWER	UPPER
Physical vs. Evaluation	0.074	0.950	-0.107	0.256
Physical vs. Money Loss	0.538	0.950	0.194	0.883
Evaluation vs. Money Loss	0.464	0.950	0.116	0.812

As you can see in Output 8.4, the confidence intervals for the physical versus money loss and evaluation versus money loss comparisons do not contain the value zero. The confidence interval for the physical versus evaluation comparison does contain the value zero. Note that the Scheffe confidence intervals are narrower than the Bonferroni confidence intervals because $c' = \min(c, p-1) = \min(3, 3-1) = 2$ is less than $c = 3$.

A planned comparison also was conducted to compare the physical and evaluation aggression measures with the money-loss aggression measure. The coefficients for this panned contrast are $c_1 = c_2 = 0.5$ and $c_3 = -1$. You can easily alter the SAS code that produces Output 8.3 to conduct planned contrasts. Simply create a SAS data set called COEFF that has one line of data that contains the three contrast coefficients (that is, 0.5 0.5 −1 in this example). You must also change the value of *kind* to 3 (that is, *kind*=3). Output 8.5 shows the results obtained from the modified SAS code.

Output 8.5 *Planned comparison using the average of physical and evaluation versus money loss*

TYPE	CONTRAST	LEVEL	LOWER	UPPER
Planned Comparison	0.501	0.950	0.209	0.793

As you can see in Output 8.5, the contrast was significant at the .05 level because the confidence interval does not include the value zero. Sex differences in physical and evaluation measures of aggression were different from sex differences in the money-loss measure of aggression.

8.4.2 Fixed-Effects ANOVA Models with Two Categorical Factors

In a meta-analysis, suppose that there are two categorical factors A and B, where A has a distinct values and B has b distinct values. Let AB_{ij} denote the group exposed to the *i*th level of A and *j*th level of B. Table 8.5 depicts the $a \times b$ separate groups in the design.

Table 8.4 Groups in a two-way ANOVA

		Factor A levels			
		1	2	...	a
Factor B levels	1	AB_{11}	AB_{21}	...	AB_{a1}
	2	AB_{12}	AB_{22}	...	AB_{a2}
	⋮	⋮	⋮	⋱	⋮
	b	AB_{1b}	AB_{2b}	...	AB_{ab}

Let θ_{ijk} denote the kth effect-size parameter at ith level of A and jth level of B, and let $\hat{\theta}_{ijk}$ denote its estimate with sampling variance $\mathrm{Var}(\hat{\theta}_{ijk})$. Table 8.5 gives the notation for a fixed-effects model with two categorical factors.

Table 8.5 *Effect-size estimates and sampling variance for $a \times b$ groups of studies*

	Effect-size estimates	Variance estimates
Group AB_{11}		
Study 1	$\hat{\theta}_{111}$	$\mathrm{Var}(\hat{\theta}_{111})$
Study 2	$\hat{\theta}_{112}$	$\mathrm{Var}(\hat{\theta}_{112})$
$\vdots$	$\vdots$	$\vdots$
Study m_{11}	$\hat{\theta}_{11m_{11}}$	$\mathrm{Var}(\hat{\theta}_{11m_{11}})$
$\vdots$	$\vdots$	$\vdots$
Group AB_{a1}		
Study 1	$\hat{\theta}_{a11}$	$\mathrm{Var}(\hat{\theta}_{a11})$
Study 2	$\hat{\theta}_{a12}$	$\mathrm{Var}(\hat{\theta}_{a12})$
$\vdots$	$\vdots$	$\vdots$
Study m_{a1}	$\hat{\theta}_{a1m_{a1}}$	$\mathrm{Var}(\hat{\theta}_{a1m_{a1}})$
$\vdots$	$\vdots$	$\vdots$
Group AB_{1b}		
Study 1	$\hat{\theta}_{1b1}$	$\mathrm{Var}(\hat{\theta}_{1b1})$
Study 2	$\hat{\theta}_{1b2}$	$\mathrm{Var}(\hat{\theta}_{1b_2})$
$\vdots$	$\vdots$	$\vdots$
Study m_{1b}	$\hat{\theta}_{1bm_{1b}}$	$\mathrm{Var}(\hat{\theta}_{1bm_{1b}})$
$\vdots$	$\vdots$	$\vdots$
Group AB_{ab}		
Study 1	$\hat{\theta}_{ab1}$	$\mathrm{Var}(\hat{\theta}_{ab_1})$
Study 2	$\hat{\theta}_{ab2}$	$\mathrm{Var}(\hat{\theta}_{ab2})$
$\vdots$	$\vdots$	$\vdots$
Study m_{ab}	$\hat{\theta}_{abm_{ab}}$	$\mathrm{Var}(\hat{\theta}_{abm_{ab}})$

In the analysis of variance for the experiment with two crossed factors and an interaction term, the total sum of squares can be partitioned into four components: (a) the sum of squares between levels of factor A, (b) the sum of squares between levels of factor B, (c) the sum of squares for the interaction AB, and (d) the sum of squares within groups (that is, error). In symbols,

$$SS_{TOTAL} = SS_A + SS_B + SS_{AB} + SS_{WITHIN}. \tag{8.20}$$

Similarly, in the two-way analysis of variance for a meta-analysis, the total heterogeneity statistic can be partitioned into four components: (a) heterogeneity between levels of factor A, (b) heterogeneity between levels of factor B, (c) heterogeneity due to the interaction AB, and (d) heterogeneity within groups of studies. In symbols,

$$Q_{TOTAL} = Q_A + Q_B + Q_{AB} + Q_{WITHIN}. \tag{8.21}$$

To test the null hypothesis that there is no variation in factor level mean-effect sizes across levels of factor A,

$$H_0 : \theta_{1++} = \cdots = \theta_{a++}, \tag{8.22}$$

you use the statistic

$$Q_A = \sum_{i=1}^{a} w_{i++} \left(\hat{\theta}_{i++} - \hat{\theta}_{+++} \right)^2, \tag{8.23}$$

where $w_{i++} = \sum_{j=1}^{b} \sum_{k=1}^{m_{ij}} w_{ijk}$, the weight $w_{ijk} = 1/\mathrm{Var}\left(\hat{\theta}_{ijk}\right)$,

$$\hat{\theta}_{i++} = \frac{\sum_{j=1}^{b} \sum_{k=1}^{m_{ij}} w_{ijk} \hat{\theta}_{ijk}}{\sum_{j=1}^{b} \sum_{k=1}^{m_{ij}} w_{ijk}} \tag{8.24}$$

is the weighted mean of the effect-size estimates in the ith level of factor A, and

$$\hat{\theta}_{+++} = \frac{\sum\limits_{i=1}^{a}\sum\limits_{j=1}^{b}\sum\limits_{k=1}^{m_{ij}} w_{ijk}\hat{\theta}_{ijk}}{\sum\limits_{i=1}^{a}\sum\limits_{j=1}^{b}\sum\limits_{k=1}^{m_{ij}} w_{ijk}\hat{\theta}_{ijk}} = \frac{\sum\limits_{i=1}^{a} w_{i++}\hat{\theta}_{i++}}{\sum\limits_{i=1}^{a} w_{i++}}, \tag{8.25}$$

is the grand weighted mean (Hedges, 1994, p. 294). You reject the null hypothesis in Equation 8.22 at significance level α if Q_A exceeds the $100(1-\alpha)\%$ critical value of the chi-square distribution with $a-1$ degrees of freedom.

To test the null hypothesis that there is no variation in factor level mean-effect sizes across levels of factor B,

$$H_0 : \theta_{+1+} = \cdots = \theta_{+b+}, \tag{8.26}$$

you use the statistic

$$Q_B = \sum_{i=1}^{a} w_{+j+}\left(\hat{\theta}_{+j+} - \hat{\theta}_{+++}\right)^2, \tag{8.27}$$

where $w_{+j+} = \sum\limits_{i=1}^{a}\sum\limits_{k=1}^{m_{ij}} w_{ijk}$ and

$$\hat{\theta}_{+j+} = \frac{\sum\limits_{i=1}^{a}\sum\limits_{k=1}^{m_{ij}} w_{ijk}\hat{\theta}_{ijk}}{\sum\limits_{i=1}^{a}\sum\limits_{k=1}^{m_{ij}} w_{ijk}} \tag{8.28}$$

(Hedges, 1994, p. 294). You reject the null hypothesis in Equation 8.26 at significance level α if Q_B exceeds the $100(1-\alpha)\%$ critical value of the chi-square distribution with $b-1$ degrees of freedom.

To test the null hypothesis that there is no AB interaction,

$$H_0 : \begin{matrix} \theta_{11+} = & \cdots & = \theta_{1b+} \\ \vdots & \ddots & \vdots \\ \theta_{a1+} = & \cdots & = \theta_{ab+} \end{matrix} \quad , \tag{8.29}$$

you use the statistic

$$Q_{AB} = \sum_{i=1}^{a} \sum_{j=1}^{b} w_{ij+} \left(\hat{\theta}_{ij+} - \hat{\theta}_{+++} \right)^2 , \tag{8.30}$$

where $w_{ij+} = \sum_{k=1}^{m_{ij}} w_{ijk}$ and

$$\hat{\theta}_{ij+} = \frac{\sum_{k=1}^{m_{ij}} w_{ijk} \hat{\theta}_{ijk}}{\sum_{k=1}^{m_{ij}} w_{ijk}} \tag{8.31}$$

(Hedges, 1994, p. 294). You reject the null hypothesis in Equation 8.29 at significance level α if Q_{AB} exceeds the $100(1-\alpha)\%$ critical value of the chi-square distribution with $(a-1)(b-1)$ degrees of freedom.

To test the null hypothesis that there is no variation in population effects within the cells that are defined by crossing the levels of factors A with the levels of factor B,

$$H_o : \theta_{ijk} = \theta_{ij+} , \tag{8.32}$$

you use the statistic

$$Q_{WITHIN} = \sum_{i=1}^{a} \sum_{j=1}^{b} \sum_{k=1}^{m_{ij}} w_{ijk} \left(\hat{\theta}_{ijk} - \hat{\theta}_{ij+} \right)^2 . \tag{8.33}$$

You reject the null hypothesis in Equation 8.32 at significance level α if Q_{WITHIN} exceeds the $100(1-\alpha)\%$ critical value of the chi-square distribution with

$\sum_{i=1}^{a} \sum_{j=1}^{b} (m_{ij} - 1)$ degrees of freedom.

Table 8.6 summarizes the various sources of heterogeneity in a meta-analysis with two categorical study characteristics, as well as the test statistics and corresponding degrees of freedom for each source of heterogeneity.

Table 8.6 *Heterogeneity summary table for a fixed-effects model with two categorical factors*

Source	Q Statistic	df
Factor A	Q_A	$a-1$
Factor B	Q_B	$b-1$
AB interaction	Q_{AB}	$(a-1)(b-1)$
Within groups	Q_{WITHIN}	$\sum_{i=1}^{a}\sum_{j=1}^{b}(m_{ij}-1)$
Within group AB_{11}	$Q_{WITHIN\,AB_{11}}$	$m_{11}-1$
$\vdots$	$\vdots$	$\vdots$
Within group AB_{a1}	$Q_{WITHIN\,AB_{a1}}$	$m_{a1}-1$
$\vdots$	$\vdots$	$\vdots$
Within group AB_{1b}	$Q_{WITHIN\,AB_{1b}}$	$m_{1b}-1$
$\vdots$	$\vdots$	$\vdots$
Within group AB_{ab}	$Q_{WITHIN\,AB_{ab}}$	$m_{ab}-1$
Total	Q_{TOTAL}	$k-1$

Note: $k = \sum_{i=1}^{a}\sum_{j=1}^{b}(m_{ij}-1)$. *df* = degrees of freedom

---

Example 8.3 *Heterogeneity tests for a two-factor model*

In this example, we use the data in Table 8.3. Factor *A* is the type of aggression that is measured (physical, evaluation, money loss) and Factor *B* is provocation (provoked versus neutral). The SAS code, along with the SAS macro in Appendix 8.1 for obtaining the various *Q* statistics, is given below. In this example, *ftname* was TRT and it had six levels (three types of aggression measures × 2 levels of provocation).

The statement `%let varlist1 = aggr prov` assigns the macro variable VARLIST1 the values AGGR and PROV for the CLASS statement in the GLM procedure. The statement `%let varlist2 = aggr prov aggr*prov` assigns the macro variable VARLIST2 the values AGGR, PROV, and AGGR*PROV for the MODEL statement in the GLM. AGGR and PROV are the main effects for type of aggression measure and provocation, respectively, and AGGR*PROV is the interaction effect.

The results from the following SAS code are printed in Output 8.6:

```
options nodate nocenter pagesize=54 linesize=80 pageno=1;
libname ch8 "d:\metabook\ch8\dataset";
data ch8.ex83;
   set ch8.ex81;
   if (aggr = 1) then trt=aggr+prov-1;
   else if (aggr = 2) then trt=aggr+prov;
   else if (aggr = 3) then trt=aggr+prov+1;
%let varlist1 = aggr prov;
%let varlist2 = aggr prov aggr*prov;
%within(ch8.ex83,ch8.ex83out,eff,veff,trt,6);
proc format;
   value aa  1 = "Aggression              "
             2 = "Provocation             ",
             3 = "Aggression*Provocation  "
             4 = "Within                  "
             5 = "  Physical and Provoked "
             6 = "  Physical and Neutral  "
             7 = "  Evaluation and Provoked"
             8 = "  Evaluation and Neutral "
             9 = "  Money Loss and Provoked"
            10 = "  Money Loss and Neutral "
            11 = "Corrected Total         ";
proc print data=ch8.ex83out noobs;
   title;
   format source aa. qstat pvalue 7.3;
quit;
```

Output 8.6 *Two-way ANOVA table for heterogeneity summary statistics*

SOURCE	DF	QSTAT	PVALUE
Aggression	2	12.296	0.002
Provocation	1	0.390	0.532
Aggression*Provocation	2	4.797	0.091
Within	58	123.202	0.000
Physical and Provoked	16	23.207	0.108
Physical and Neutral	16	45.797	0.000
Evaluation and Provoked	12	24.905	0.015
Evaluation and Neutral	12	28.314	0.005
Money Loss and Provoked	1	0.941	0.332
Money Loss and Neutral	1	0.038	0.846
Corrected Total	63	140.685	0.000

As you can see in Output 8.6, there was a main effect for type of aggression measure. This main effect was interpreted in Example 8.1. The main effect for provocation and the interaction between type of aggression measure and provocation were both nonsignificant. Note that the within-groups heterogeneity statistic is still significant, even when two factors are included in the model. Three of the six groups were heterogeneous (that is, Physical and Neutral, Evaluation and Provoked, and Evaluation and Neutral). Because the "extra" variation in effect-size estimates cannot be explained by the two factors, a random-effects model might be more appropriate than a fixed-effects model for this data set.

8.5 Testing the Moderating Effects of Continuous Study Characteristics in Regression Models

You can use regression analysis to determine the relation between continuous study characteristics and effect-size estimates (for example, Hedges & Olkin, 1985). Some examples of continuous study characteristics include length of exposure to a treatment, publication year, and age of participants. You also can use regression analysis to determine the relation between categorical study characteristics and effect-size estimates using dummy variables (that is, variables coded 1 or 0). If a study characteristic has p levels, you need $p - 1$ dummy variables. For dichotomous study characteristics, only one indicator variable is needed. For example, you could code published studies 1 and unpublished studies 0. If a study characteristic has several levels, however, ANOVA models are less complicated because there are fewer variables in the model.

Let $\hat{\theta}_1, \ldots, \hat{\theta}_k$ be k independent effect-size parameters, and let $\hat{\theta}_1, \ldots, \hat{\theta}_k$ be the corresponding effect-size estimates with sampling variances $\mathrm{Var}(\hat{\theta}_1), \ldots, \mathrm{Var}(\hat{\theta}_k)$. The fixed-effects regression model is given by

$$\theta_i = \beta_0 + \beta_1 X_{i1} + \ldots + \beta_p X_{ip}, \tag{8.34}$$

where β_0 is the intercept and $\beta_1, \ldots, \beta_p$ are the unstandardized regression coefficients, and $X_1, \ldots, X_p$ are the p known study characteristics (Hedges, 1994, p. 295). The weighted least squares estimates of the regression parameters $\beta_0, \beta_1, \ldots, \beta_p$ are $b_0, b_1, \ldots, b_p$, where the weight for the effect-size estimate $\hat{\theta}_i$ is defined as $w_i = 1 / \mathrm{Var}(\hat{\theta}_i)$.

8.5.1 Confidence Intervals for Individual Regression Coefficients

Suppose that the weighted least estimates $b_0, b_1, \ldots, b_p$, are normally distributed about their respective parameters $\beta_0, \beta_1, \ldots, \beta_p$ with estimated variances $\mathrm{Var}(b_0), \mathrm{Var}(b_1), \ldots, \mathrm{Var}(b_p)$. The $100(1 - \alpha)\%$ confidence interval for the individual regression coefficient β_j is given by

$$b_j - Z_{\alpha/2} \frac{\sqrt{\mathrm{Var}(b_j)}}{\sqrt{MS_{ERROR}}} \leq \beta_j \leq b_j + Z_{\alpha/2} \frac{\sqrt{\mathrm{Var}(b_j)}}{\sqrt{MS_{ERROR}}}, \tag{8.35}$$

where MS_{ERROR} is the error mean square for the regression model (Hedges, 1994, p. 296).

Example 8.4 *Estimates and confidence intervals for individual regression coefficients*

Example 8.4 comes from a meta-analysis on the relation between group cohesiveness and performance (Mullen & Copper, 1994). Researchers typically measure group cohesiveness in terms of interpersonal attraction, group pride, and commitment to task performance. It is commonly assumed that cohesive groups perform better on tasks than do noncohesive groups. Mullen and Cooper (1994) tested this hypothesis in their meta-analysis. They also coded study characteristics that might moderate the relation between group cohesiveness and performance, including group size, whether the group was real or artificial, whether the group required high or low interaction, and whether the study was correlational or experimental. The relation between group cohesiveness and performance was expected to be stronger for smaller groups, for real groups, for groups that required members to interact, and for correlational studies (because they are generally more realistic than experimental studies). The data for the meta-analysis are given in Table 8.7.

Table 8.7 *Mullen and Cooper (1994) meta-analysis of the relation between group cohesiveness and performance*

Study	N	r	Group Size	Reality	Paradigm	Interaction
1	48	.645	5.0	1	1	0
2	40	.392	3.0	0	0	1
3	40	.316	3.0	0	0	1
4	40	.478	3.0	0	0	1
5	431	.140	8.8	1	1	1
6	86	.086	3.9	0	1	1
7	183	.680	6.0	1	1	1
8	200	.320	8.8	1	1	0
9	32	.018	4.0	0	0	1
10	281	.188	6.2	1	1	1
11	130	.400	10.0	1	1	1
12	93	-.170	10.3	0	1	1
13	93	.326	3.5	0	1	1
14	275	.158	12.0	1	1	1
15	72	.316	4.0	0	1	1
16	231	.100	11.0	1	1	1
17	370	.040	11.2	1	1	0
18	72	.770	6.0	1	1	1
19	259	.194	10.0	1	1	1

(continued)

Study	N	r	Group Size	Reality	Paradigm	Interaction
20	41	.098	6.5	1	1	0
21	178	.300	5.0	1	1	1
22	702	.051	13.0	1	1	1
23	16	.290	4.0	0	1	1
24	450	.180	5.0	1	1	1
25	64	.370	4.0	0	0	1
26	64	.497	4.0	0	0	1
27	64	-.172	4.0	0	0	1
28	62	.431	4.0	0	0	1
29	62	.125	4.0	0	0	1
30	64	.101	4.0	0	0	1
31	110	.157	5.5	0	1	1
32	278	.470	8.7	1	1	1
33	71	.762	5.0	1	1	1
34	495	.190	9.0	1	1	1
35	147	.469	7.0	1	1	1
36	53	.201	3.0	1	1	0
37	135	.063	5.0	1	1	1
38	112	.120	6.2	1	1	1
39	68	.440	3.8	0	1	1
40	40	.212	5.0	0	1	1
41	80	.490	5.5	1	1	1
42	263	.178	8.2	1	1	0
43	90	.415	5.0	0	0	0
44	100	.200	20.0	1	1	1
45	86	.075	4.0	0	1	1
46	69	-.040	8.8	1	1	1
47	125	.049	5.0	0	1	1
48	176	.560	5.0	1	1	1
49	133	-.039	3.2	0	1	1
50	345	.320	3.0	1	1	1
51	66	.440	3.0	1	1	1
52	132	.811	11.0	1	1	1
53	83	.409	4.6	1	1	0
54	162	.320	4.5	0	1	1
55	28	.417	3.0	0	0	0
56	26	.052	3.0	0	0	0
57	28	.307	3.0	0	0	0
58	28	.134	3.0	0	0	0
59	28	.250	3.0	0	0	0
60	26	.362	3.0	0	0	0
61	64	.270	4.0	0	0	1
62	64	.101	4.0	0	0	1
63	68	.190	4.0	0	0	1
64	68	-.095	4.0	0	0	1
65	68	.088	4.0	0	0	1
66	64	.176	4.0	0	0	1

Note: Reality: 1 = real group, 0 = artificial group; Paradigm: 1 = correlational,
0 = experimental; Interaction: 1 = high interaction required, 0 = low interaction required.

The data set CH8.EX84, which can be accessed via SAS Online Samples (see "Using This Book"), contains the data in Table 8.9. The data set contains six variables: (a) a continuous variable GROUP for group size, (b) a dummy variable REALITY denoting whether the group was real (coded 1) or artificial (coded 0), (c) a dummy variable PARADIGM denoting whether the study was correlational (coded 1) or experimental (coded 0), (d) a dummy variable INTERACT denoting whether high (coded 1) or low (coded 0) interaction was required of group members, (e) a continuous variable EFF for the effect-size estimate (the correlation coefficient in this example), and (f) a continuous variable VEFF for the estimated variance of the effect-size estimate.

You can use the SAS macro FIXPARA(*indata,outdata,level,eff,veff*) in Appendix 8.4 to obtain confidence intervals for study characteristics in fixed-effects regression models. *Indata* and *outdata* are the names of the input and output data sets, respectively. *Level* is the level of confidence for the confidence interval (for example, .95 for a 95% confidence interval). *Eff* is the effect-size estimate, and *veff* is the estimated variance of *eff*. If the estimated effect-size variances are not reported, but the sample sizes are reported, you can obtain the variances using the SAS macros in Chapter 5, Appendix 5.3. If the effect-size estimate is very small, you may want to change the format statement in the second-to-last line of the SAS code that follows. For example, if the estimate is 0.00000123, the value 6.3 will give the estimate to six digits (that is, 0.00000). Replacing the value 6.3 with 9.3 gives the more useful estimate 0.00000123.

```
options nodate nocenter pagesize=54 linesize=80 pageno=1;
libname ch8 "d:\metabook\ch8\dataset";
%fixreg(ch8.ex84,ch8.ex84out,.95,eff,veff);
proc format;
   value aa 1 = "Intercept"
            2 = "Group    "
            3 = "Reality  "
            4 = "Paradigm "
            5 = "Interact ";
proc print data=ch8.ex84out noobs;
   title;
   format param aa. estimate lower upper 6.3;
run;
```

Output 8.7 Confidence intervals for individual regression coefficients in the meta-analysis on group cohesiveness and group performance

PARAM	ESTIMATE	LEVEL	LOWER	UPPER
Intercept	0.303	0.95	0.221	0.384
Group	-0.027	0.95	-0.035	-0.020
Reality	0.224	0.95	0.153	0.295
Paradigm	-0.059	0.95	-0.144	0.027
Interact	0.043	0.95	-0.018	0.103

As expected, the smaller the group, the stronger the relation between group cohesiveness and performance. Real groups also were more cohesive than artificial groups (variable REALITY). The other study characteristics (that is, high versus low interaction required among group members, variable INTERACT; experimental versus correlational design, variable PARADIGM) were nonsignificant.

8.5.2 Omnibus Tests for Blocks of Regression Coefficients and Tests for Homogeneity of Effects

You may want to test hypotheses about groups or blocks of regression coefficients. Using Hedges' (1994) notation, suppose that a block of a study characteristics $X_1, ..., X_a$ has already been entered in the regression equation and you want to test whether a second block of b additional study characteristics $X_{a+1}, ..., X_{a+b}$ explains any of the variation in effect sizes not explained by the first block. To test the null hypothesis

$$H_0 : \beta_{a+1} = ... = \beta_{a+b} = 0, \qquad (8.36)$$

use the test statistic

$$Q_{CHANGE} = b \times F_{CHANGE} \times MS_{ERROR}, \qquad (8.37)$$

where F_{CHANGE} is the value of the F test for the significance of the addition of the block of b study characteristics and MS_{ERROR} is the weighted error mean square from the regression analysis with $a+b$ predictor variables (Hedges, 1994, p. 297).

You reject the null hypothesis in Equation 8.36 at significance level α if Q_{CHANGE} exceeds the $100(1-\alpha)\%$ critical value of the chi-square distribution with b degrees of freedom.

You use the Q_{CHANGE} statistic to determine whether the entire block of p study characteristics explains a significant amount of variation in effect sizes. In this case, a equals 0, b equals p. You reject the null hypothesis

$$H_0 : \beta_1 = \ldots = \beta_p = 0, \tag{8.38}$$

if the test statistic

$$
\begin{aligned}
Q_{CHANGE} &= b \times F_{CHANGE} \times MS_{ERROR} \\
&= p \times \frac{Q_{MODEL}/p}{Q_{ERROR}/(k-p-1)} \times \frac{Q_{ERROR}}{k-p-1} \\
&= Q_{MODEL}
\end{aligned}
\tag{8.39}
$$

exceeds the $100(1-\alpha)\%$ critical value of the chi-square distribution with $b = p$ degrees of freedom.

Similarly, suppose that a block of $p-1$ study characteristics has already been entered in the regression model and you want to know whether the study characteristic X_i explains any additional variation in effect sizes. That is, you want to know whether a given study characteristic explains additional variation in effect sizes after controlling for all other study characteristics. In this case, $a = p-1$ and $b = 1$. You reject the null hypothesis

$$H_0 : \beta_i = 0 \tag{8.40}$$

if the test statistic

$$
\begin{aligned}
Q_{CHANGE} &= b \times F_{CHANGE} \times MS_{ERROR} \\
&= 1 \times \frac{Q_i/1}{Q_{ERROR}/(k-p-1)} \times \frac{Q_{ERROR}}{k-p-1} \\
&= Q_i
\end{aligned}
\tag{8.41}
$$

exceeds the $100(1-\alpha)\%$ critical value of the chi-square distribution with $b=1$ degree of freedom.

If k, the number of effect-size estimates, is greater than $(p + 1)$, the number of study characteristics plus the intercept, you can conduct the heterogeneity of effect sizes test after controlling for the p study characteristics using the weighted error sum of squares Q_{ERROR} (Hedges, 1994, p. 297). You reject the null hypothesis that the effect sizes are homogeneous at level α, after controlling for the p study characteristics, if Q_{ERROR} exceeds the $100(1-\alpha)\%$ critical value of the chi-square distribution with $(k - p - 1)$ degrees of freedom.

Example 8.5 *Omnibus tests for blocks of regression coefficients in the meta-analysis on group cohesion and performance*

Suppose that you are interested in testing whether all study characteristics in the meta-analysis on group cohesion and performance explain a significant amount of variation in effect sizes. The SAS code for conducting this test is given next, and the results are printed in Output 8.8. Note that the model in PROC GLM contains all four study characteristics.

```
options nodate nocenter pagesize=54 linesize=80 pageno=1;
libname ch8 "d:\metabook\ch8\dataset";
data  bb;
   set ch8.ex84;
   weight = 1 / veff;
proc glm data=bb noprint outstat=aa;
   weight weight;
   model eff = group reality paradigm interact;
data cc dd;
   set aa;
   qstat=ss;
   if (_type_ = "ERROR") then output dd;
   if (_type_ = "SS1  ") then output cc;
   keep df qstat;
proc means data=cc noprint;
   var df qstat;
   output out=bb sum=df qstat;
data ch8.ex85out;
   source = _n_;
   set bb cc dd;
   pvalue=1-probchi(qstat,df);
   keep source df qstat pvalue;
```

```
proc format;
   value aa 1 = "Model      "
            2 = "  Group    "
            3 = "  Reality "
            4 = "  Paradigm"
            5 = "  Interact"
            6 = "Error      ";
proc print noobs;
   title;
   format source aa. qstat pvalue 8.4;
quit;
```

Output 8.8 *ANOVA table for regression analysis of studies of the relation between group cohesiveness and group performance*

SOURCE	DF	QSTAT	PVALUE
Model	4	69.2727	0.0000
Group	1	22.3861	0.0000
Reality	1	43.6468	0.0000
Paradigm	1	1.3286	0.2491
Interact	1	1.9113	0.1668
Error	61	420.1070	0.0000

As you can see in Output 8.8, the overall model explained a significant amount of variation in effect sizes. The study characteristics GROUP (that is, group size) and REALITY (that is, whether the group was real or artificial) both explained a significant amount of variation in effect sizes. However, the study characteristics PARADIGM (that is, whether the study was correlational or experimental) and INTERACT (that is, whether high or low interaction was required of group members) did not explain a significant amount of variation in effect sizes. The Q_{ERROR} statistic is also significant, suggesting that the effect-size estimates are too heterogeneous to be combined.

To explore whether the study characteristics PARADIGM and INTERACT should be excluded from the regression model, a second analysis was conducted. In this second analysis, GROUP and REAL were entered as the first block of variables, and PARADIGM and INTERACT were entered as the second block of variables. The following SAS code was used to produce the results in Output 8.9:

```
options nodate nocenter pagesize=54 linesize=80 pageno=1;
libname ch8 "d:\metabook\ch8\dataset";
data aa bb;
    set ch8.ex85out;
    if (source = 4 or source = 5) then output aa;
    if (source > 3) then output bb;
    keep df qstat;
proc means data=aa;
    var df qstat;
    output out=cc sum=df qstat;
proc means data=bb;
    var df qstat;
    output out=dd sum=df qstat;
data;
    source=_n_;
    set cc dd;
    pvalue=1-probchi(qstat,df);
    keep source df qstat pvalue;
proc format;
    value aa 1 = "QCHANGE"
             2 = "QERROR ";
proc print noobs;
    title;
    format source aa. qstat pvalue 7.3;
run;
```

Output 8.9 *Omnibus test for the study characteristics PARADIGM and INTERACT after controlling for the study characteristics REALITY and GROUP in the meta-analysis on group cohesion and performance*

SOURCE	DF	QSTAT	PVALUE
QCHANGE	2	3.240	0.198
QERROR	63	423.347	0.000

Because the p-value for Q_{CHANGE} is greater than .05, you fail to reject the null hypothesis that the PARADIGM and INTERACT study characteristics are not related to effect sizes after controlling for the GROUP and REALITY study characteristics. Thus, you probably should exclude the study characteristics PARADIGM and INTERACT from the regression model. The Q_{ERROR} statistic is also significant, suggesting that the effect-size estimates are too heterogeneous to be combined.

Finally, we perform a regression analysis that only includes the two study characteristics with significant regression coefficients (that is, group size and whether the group was real or artificial). The following SAS code produced the results in Output 8.10:

```
options nodate nocenter pagesize=54 linesize=80 pageno=1;
libname ch8 "d:\metabook\ch8\dataset";
data  bb;
   set ch8.ex84;
   weight = 1 / veff;
proc glm data=bb noprint outstat=aa;
   weight weight;
   model eff = group reality;
data cc dd;
   set aa;
   qstat=ss;
   if (_type_ = "ERROR") then output dd;
   if (_type_ = "SS1  ") then output cc;
   keep df qstat;
proc means data=cc noprint;
   var df qstat;
   output out=bb sum=df qstat;
data ch8.ex85out3;
   source = _n_;
   set bb cc dd;
   pvalue=1-probchi(qstat, df);
   keep source df qstat pvalue;
proc format;
   value aa 1 = "Model    "
            2 = "  Group  "
            3 = "  Reality"
            4 = "Error    ";
proc print noobs;
   title;
   format source aa. qstat pvalue 8.4;
quit;
```

Output 8.10 *ANOVA table for regression analysis of studies of the relation between group cohesiveness and group performance with only two variables*

SOURCE	DF	QSTAT	PVALUE
Model	2	66.0329	0.0000
Group	1	22.3861	0.0000
Reality	1	43.6468	0.0000
Error	63	423.3469	0.0000

As you can see in Output 8.10, although both study characteristics are significant, the Q_{ERROR} statistic is still significant, indicating that the effect-size estimates are heterogeneous. Because the "extra" variation in effect-size estimates cannot be explained using a few study characteristics, a random-effects model may be more appropriate than a fixed-effects model for this data set (see Chapter 9 for a discussion of random-effects models in meta-analysis.)

8.5.3 Multicollinearity among Study Characteristics

Multicollinearity, or correlation among the predictor variables, can be a problem whenever multiple regression analysis is used in an individual study or in a meta-analysis. In a meta-analysis, study characteristics tend to be correlated in studies from the same laboratory. If multicollinearity exists among study characteristics, the reviewer cannot determine the effect of each study characteristic on effect-size estimates because the effect of each study characteristic depends upon which other study characteristics are included in the regression model.

Multicollinearity can be assessed by means of variance inflation factors, VIF (see Neter, Wasserman, & Kutner, 1990). A VIF of 1 indicates that the study characteristics are not linearly related. A maximum VIF value in excess of 10 is often taken as an indication that multicollinearity may be unduly influencing the least squares estimates.

You can compute a VIF for each study characteristic in the regression model using the following formula:

$$VIF_i = \frac{1}{1 - (SS3_i / SS_{TOTAL})},$$ (8.42)

where $SS3_i$ is the Type III sum of squares for ith study characteristic and SS_{TOTAL} is the corrected total sum of squares for the weighted least squares regression model. You can obtain both sums of squares from the OUTSTAT option in PROC GLM.

Example 8.6 *Testing for multicollinearity among study characteristics in the meta-analysis on group cohesion and performance*

The following SAS code computes the VIF for each study characteristic in the meta-analysis on group cohesion and performance (Mullen & Cooper, 1994). The results are printed in Output 8.11.

```
options nodate nocenter pagesize=54 linesize=80 pageno=1;
libname ch8 "d:\metabook\ch8\dataset";
data;
   set ch8.ex84;
   weight = 1 / veff;
proc glm noprint outstat=aa;
   weight weight;
   model eff = group reality paradigm interact;
data;
   set aa;
   if (_type_ ^= "SS3   ");
proc means noprint;
   var ss;
   output out=bb sum=ssct;
data ch8.ex86;
   set aa;
   source=_n_-5;
   if (_type_ = "SS3   ");
   keep source ss;
data;
   set ch8.ex86;
   if (_n_ = 1) then set bb;
   vif = 1 / (1-ss/ssct);
   keep source vif;
proc format;
   value aa 1 = "Group   "
           2 = "Reality "
           3 = "Paradigm"
           4 = "Interact";
proc print noobs;
   format source aa. vif 5.3;
   title;
quit;
```

Output 8.11 Variance inflation factors for study characteristics in the meta-analysis on group cohesion and performance

```
SOURCE          VIF

Group           1.124
Reality         1.084
Paradigm        1.004
Interact        1.004
```

As you can see in Output 8.11, the maximum VIF is 1.124, a value only slightly larger than 1 and much smaller than 10. Thus, multicollinearity is not a problem in this example.

8.6 Quantifying Variation Explained by Study Characteristics in ANOVA and Regression Models

Although heterogeneity statistics are useful in testing whether the unexplained variation in effect-size estimates is greater than what you would expect by chance alone, they are not useful in quantifying the amount of unexplained variation. The Birge ratio, R_B, is often used in the physical sciences to quantify the amount of unexplained variation. The Birge ratio estimates the ratio of the between-studies variation in effects to the within-studies variation in effects. Thus, a Birge ratio of 1.5 suggests that there is 50% more between-studies variation than might be expected, given the amount of within-studies variation. The value of R_B should be about 1.0 if the model fits the data exactly.

In a fixed-effects ANOVA model, the Birge ratio is defined as

$$R_B = \frac{Q_{WITHIN}}{k - p} \qquad (8.43)$$

(Hedges, 1994, p. 298). In the meta-analysis on sex differences in aggression, the Birge ratio is 2.10 (that is, 128.390/61) for the one-factor ANOVA model and 2.12 (that is, 123.2/58) for the two-factor ANOVA model. In both models, there is about two times more between-studies variation than you might expect, given the amount of within-studies variation.

In a fixed-effects regression model with an intercept, the Birge ratio is defined as

$$R_B = \frac{Q_{ERROR}}{k - p - 1}$$

(8.44)

(Hedges, 1994, p. 298). Note that the Birge ratio is simply the error mean squares of the weight least squares regression model (8.34). In the meta-analysis on group cohesiveness and performance, the Birge ratio is 6.887 (420.107/61). This means that there is about seven times more between-studies variation than you might expect, given the amount of within-studies variation. The maximum squared multiple correlation for this example is about 85 percent (that is, (6.89–1)/6.89). The between-study variation explained by our model is about 14 percent (that is, model sum of squares divided by corrected total sum of squares = 69.27/489.38). Thus, this model explains about 16 percent (that is, 14%/86%) of the explainable between-studies variation.

8.7 Conclusions

Fixed-effects procedures in meta-analysis are analogous to fixed-effects procedures in primary research. This chapter considered both categorical and continuous study characteristics. You can test the influence of categorical study characteristics on effect-size estimates using ANOVA. This chapter described both one-way and two-way ANOVA procedures. If a study characteristic is related to effect-size estimates, and the study characteristic contains more than two groups, you can perform *a priori* or *post hoc* comparisons to decide how the groups differ. You can use regression analysis to test the influence of continuous study characteristics on effect-size estimates. If you use dummy variables, you also can use regression analysis to test categorical study characteristics.

This chapter described how to test the significance of regression coefficients, both individually and in blocks. If multiple study characteristics are included in a regression model, you should also test whether multicollinearity is a problem. The presence of multicollinearity limits your ability to draw conclusions about the individual predictors in the model. The Birge ratio was presented as a descriptive statistic for quantifying the amount of variation not explained by study characteristics.

You should use fixed-effects models in meta-analysis if the variation in study effects can be captured by a few simple study characteristics. If the variation in study effects is too complicated to be captured by a few simple study characteristics, then you should use a random-effects model instead. We recommend the use of both fixed- and random-effects models on the same data set. A sensitivity analysis can then focus on whether the type of model that you used influenced the type of conclusions drawn. Random-effects models in meta-analysis are discussed in the next chapter.

8.8 References

Bettencourt, B. A., & Miller, N. (1996). Gender differences in aggression as a function of provocation: A meta-analysis. *Psychological Bulletin, 119,* 422–447.

Buss, A. H. (1961). The psychology of aggression. New York: Wiley.

Dowdy, S. & Weardon, S. (1983). *Statistics for research.* New York: Wiley.

Hedges, L. V. (1994). Fixed effects models. In H. Cooper & L. V. Hedges (Eds.). *The handbook of research synthesis* (pp. 285–299). New York: Russell Sage Foundation.

Hedges, L. V., & Olkin, I. (1985). *Statistical methods for meta-analysis.* New York: Academic Press.

Mullen, B., & Copper, C. (1994). The relation between group cohesiveness and performance: An integration. *Psychological Bulletin, 115,* 210–227.

Neter, J., Wasserman, W., & Kutner, M. H. (1990). *Applied linear statistical models* (3rd ed.). Homewood, IL: Irwin.

Rohsenow, D. J., & Bachorowski, J. (1984). Effect of alcohol and expectancies on verbal aggression in men and women. *Journal of Abnormal Psychology, 93,* 418–432.

Rosenthal, R. (1995). Writing meta-analytic reviews. *Psychological Bulletin, 118,* 183–192.

8.9 Appendices

Appendix 8.1: SAS Macro for Computing Q-Statistics in Fixed-Effects ANOVA Models

```
%macro within(indata,outdata,eff,veff,ftname,nlevels);
/**********************************************************/
/*   INDATA:  Input data file name                      */
/*   Input data file contains three variables:          */
/*     (1) EFF: Effect-size estimate                    */
/*     (2) VEFF: The estimated variance for the         */
/*              effect-size estimate                    */
/*     (3) ftname:  Name of the factor included in      */
/*              the model                               */
/*   NLEVELS:  The number of levels of the factor       */
/*              FTNAME.  For example, if the model       */
/*              contains two categorical variables, A and */
/*              B, where A has three levels and B has two */
/*              levels, the number of levels is 6 (2*3).  */
/*              In this example, one needs to create an  */
/*              artificial variable, FTNAME, that has six */
/*              levels.                                  */
/*   OUTDATA: The output data file name.                */
/*     The output data file contains the information in */
/*     the fixed-effects ANOVA table                    */
/**********************************************************/
data tempin;
   set &indata;
   weight = 1 / &veff;
data fate;
%do i = 1 %to &nlevels;
data in&i;
   set tempin;
   if (&ftname = &i);
proc glm data=in&i noprint outstat=temp&i;
   class &ftname;
   model &eff = &ftname;
   weight weight;
quit;
```

```
%end;
%do i = 1 %to &nlevels;
   data fate;
      set fate temp&i;
%end;
proc glm data=tempin noprint outstat=out1;
   class &vlist1;
   model &eff = &vlist2;
   weight weight;
data out0 out2;
   set out1;
   if (_type_ = "SS1  ") then output out0;
   if (_type_ ^= "SS3  ") then output out2;
   keep df ss;
data out3;
   set out1 fate;
   if (_type_ = "ERROR");
   keep df ss;
proc means data=out2 noprint;
   var df ss;
   output out=out4 sum=df ss;
data &outdata;
   source = _n_;
   set out0 out3 out4;
   qstat=ss;
   pvalue = 1 - probchi(qstat,df);
   keep source qstat df pvalue;
%mend within;
```

Appendix 8.2: SAS Macro for Computing Confidence Intervals for Group Mean Effects in Fixed-Effects ANOVA Models

```
%macro wavgeff(indata,outdata,level);
/*********************************************************/
/*   INDATA:   The input data set name                 */
/*     The macro assumes that the:                     */
/*         (1) variable name for the effect-size estimate */
/*     is EFF                                           */
/*     Note that EFF can be any of the following effect */
/*     size estimators: Hedges' g, Hedges' gu, Cohen's d,*/
/*     point-biserial correlation, or Fisher's z.       */
/*         (2) variable name for the corresponding     */
/*     variance is VEFF                                 */
/*                                                      */
/*   OUTDATA: The output data file name                */
/*   The output data file contains four variables      */
/*                                                      */
/*         (1) ESTIMATE:  Combined estimate for the effect */
/*                 size                                 */
/*         (2) LOWER: Lower bound for confidence interval */
/*         (3) UPPER: Upper bound for confidence interval */
/*         (4) LEVEL: Level of confidence for the       */
/*         confidence interval.                         */
/*                                                      */
/*     LEVEL:  Level of the confidence of the confidence */
/*             interval                                 */
/*********************************************************/
data;
   set &indata;
   weight = 1 / veff;
   keep eff veff weight &varlist;
proc sort;
   by &varlist;
proc means noprint;
   var eff;
   weight weight;
   by &varlist;
   output out=&outdata mean=meff sumwgt=invvar;
run;
data &outdata;
   set &outdata;
   estimate = meff;
   level=&level;
   variance = 1 / invvar;
   std = sqrt(1/invvar);
   lower = estimate + probit(.5-.5*level)*std;
   upper = estimate + probit(.5+.5*level)*std;
   keep &varlist estimate variance level lower upper;
 run;
 %mend wavgeff;
```

Appendix 8.3 SAS Macro for Comparing Group Mean Effects in Fixed-Effects ANOVA Models

```
%macro fixconst(indata,outdata,coeff,kind,level);
/****************************************************/
/*    Input parameters:                          */
/*    INDATA: Input data file name that contains the  */
/*      effect-size estimate and its corresponding   */
/*      variance                                  */
/*    OUTDATA: Output data file name              */
/*    COEFF:  Input data file for contrast if KIND is 3  */
/*            COEFF is empty if KIND is 1 or 2    */
/*    KIND: Type of comparison                    */
/*        KIND = 1:  Bonferroni                   */
/*        KIND = 2:  Schefee                      */
/*        KIND = 3:  A priori or planned comparison  */
/*    LEVEL: The level of confidence for the confidence  */
/*      interval                                  */
/*                                                */
/*    The output data file contains the following  */
/*        variables:                             */
/*    TYPE: Type of comparison                    */
/*    CONTRAST: Contrast estimator                */
/*    LEVEL:  Level of confidence for the confidence  */
/*            interval                            */
/*    LOWER:  Lower bound of the confidence interval  */
/*    UPPER:  Upper bound of the confidence interval  */
/****************************************************/
proc iml;
    reset nolog;
    reset storage=ch8.imlrout;
    use &indata;
    read all var{&varlist} into y;
    if (&kind =3) then do;
        use &coeff;
        read all var _num_ into x;
    end;
```

```
    start compute;
    n=nrow(y);
    if (&kind ^= 3) then do;
        x = j(n#(n-1)/2,n,0);
        k = 1;
        do i = 1 to n-1;
            do j = i+1 to n;
                x[k,j]=-1;
                x[k,i]=1;
                k=k+1;
            end;
        end;
    end;
    type=j(nrow(x),1,0);
    do i = 1 to nrow(x);
        type[i]=i;
    end;
    if (&kind = 1) then div=nrow(x);
    else if (&kind = 2) then div=min(nrow(x),n-1);
    else if (&kind = 3) then div=1;
    contrast = x*y[,1];
    tt = ((x##2)*y[,2])##.5;
    level = j(nrow(x),1,&level);
    lower = contrast+probit((1-level)/(2#div))#tt;
    upper = contrast+probit(1-(1-level)/(2#div))#tt;
    create &outdata var{type contrast level lower upper};
    append;
    close &outdata;
    finish compute;
    run compute;
reset storage=ch8.imlrout;
store module=(compute);
run;
quit;
%mend fixconst;
```

Appendix 8.4: SAS Macro for Computing Confidence Intervals for Regression Coefficients in Fixed-Effects Models

```
%macro fixreg(indata,outdata,level,eff,veff);
/**********************************************************/
/*    Input Variables:                                  */
/*      INDATA:  Input data file name                   */
/*      The input data file contains only the regression */
/*      parameters, the effect-size estimates, and       */
/*      the estimated effect-size variances              */
/*      OUTDATA: Outdata file name                       */
/*      LEVEL: The level of confidence for the           */
/*            confidence interval                        */
/*      EFF: Variable name for the effect-size estimate  */
/*      VEFF: Variable name for the estimated            */
/*            effect-size variance                       */
/*                                                       */
/*    Variables in output data file:                    */
/*      PARAM: 1=Interception                            */
/*             2=The first variable in the model         */
/*             3=The second variable in the model        */
/*             p+1= The pth variable in the model        */
/*      ESTIMATE: The regression coefficient estimates   */
/*      LEVEL: Level of confidence for the regression    */
/*            coefficient confidence intervals           */
/*      LOWER: Lower bound of the confidence interval    */
/*      UPPER: Upper bound of the confidence interval    */
/**********************************************************/
data temp;
   set &indata;
   weight=1/&veff;
   drop &veff;
proc iml;
   reset nolog;
   reset storage=ch8.imlrout;
   use temp;
   start fixpara;
   read all var _num_ into dd;
   read all var {eff} into y;
   read all var {weight} into weight;
   one=j(nrow(dd),1,1);
   x=(one || dd[,1:ncol(dd)-2]);
   w=diag(weight);
   v=inv(w);
   ri=inv(sqrt(v));
   ys=ri*y;
   xs=ri*x;
   xspxs=xs`*xs;
   xspys=xs`*ys;
   b=inv(xspxs)*xspys;          /* regression coefficients */
   yhat=x*b;                    /* predicted values */
   resid=y-yhat;                /* residuals */
   sse=(ys-xs*b)`*(ys-xs*b);    /* residual sum of squares */
```

```
    dfe=nrow(x)-ncol(x);          /* error degrees of freedom */
    mse = sse / dfe;              /* residual mean squares */
    ncx=ncol(x);
    stdb=sqrt(vecdiag(inv(xspxs)));
    level=repeat(&level245ncx);
    lower=b+probit(.5-.5#level)#stdb;
    upper=b+probit(.5+.5#level)#stdb;
                                  /* confidence interval for */
                                  /* regression coefficients */
/***************************************************/
/*  Create output data file                     */
/***************************************************/
    param=(1:ncx)`;
    estimate=b;
    create fixout1 var{param estimate level lower upper};
    append;
    close fixout1;
    finish fixpara;
    run fixpara;
    reset storage=ch8.imlrout;
    store module=(fixpara);
run;
quit;
data &outdata;
    set fixout1;
run;
%mend fixreg;
```

chapter 9

Random-Effects Models in Meta-Analysis

9.1 Introduction

You should not combine heterogeneous effect-size estimates. A statistically significant heterogeneity test implies that variation in effects between-studies is significantly larger than would be expected due to sampling variation. You can treat between-studies variation in effects as fixed or random (Hedges & Olkin, 1985). The fixed-effects model assumes that the population effect size is a single fixed value, whereas the random-effects model assumes that the population effect size is a randomly distributed variable with its own mean and variance. When between-studies effect-size variation is treated as fixed, the only source of effect-size variation treated as random is the within-studies sampling variation. The meta-analyst can try to explain the "extra" variation in effects between-studies by entering known study characteristics in an analysis of variance (ANOVA) or regression model. If the "extra" variation in effects between studies can be explained by a few simple study characteristics, then you should use a fixed-effects model. When you use a fixed-effects model you can make generalizations to a universe of studies with similar study characteristics. The primary advantage of a fixed-effects model is that it is more powerful than a random-effects model (Rosenthal, 1995). Consequently, a fixed-effects model will produce narrower confidence intervals for effect-sizes than will a random-effects model.

When the "extra" variation in effects between studies is too complicated to be explained by a few study characteristics, you should use a random-effects model instead. When you use a random-effects model, you can make generalizations to a universe of such diverse studies. However, a random-effects model is generally used only as a "last resort" by many meta-analysts because it produces wider confidence intervals for effect sizes than does a fixed-effects model.

This chapter on random-effects models is parallel to the previous chapter on fixed-effects models. Differences in effects between studies can often be explained by known study characteristics. Categorical study characteristics with more than two categories or groups can be treated as factors in ANOVA models. We consider one- and two-factor ANOVA models in this chapter. You can enter continuous study characteristics as variables in regression models. You also can use dummy variables (that is, variables coded 0 or 1) can also be used in regression models. Dummy variables are especially useful for dichotomous study characteristics (for example, published versus unpublished study). If a study characteristic contains p groups, then $p-1$ dummy variables are needed.

Let $\theta_1, \ldots, \theta_k$ be k independent effect-size parameters, and let $\hat{\theta}_1, \ldots, \hat{\theta}_k$ be the corresponding effect-size estimates with corresponding estimated variances $\text{Var}(\hat{\theta}_1) \ldots, \text{Var}(\hat{\theta}_k)$. Assume that the relation between the study characteristics and the true effect size for each study can be explained by either a fixed-effects model,

$$\theta_i = \beta_0 + \beta_1 X_{i1} + \ldots + \beta_p X_{ip} \tag{9.1}$$

or a random-effects model

$$\theta_i = \beta_0 + \beta_1 X_{i1} + \ldots + \beta_p X_{ip} + u_i, \tag{9.2}$$

where $X_{i1} \ldots, X_{ip}$ are the p study characteristics (that is, continuous or dummy variables) for ith study, β_0 is the intercept, $\beta_1, \ldots, \beta_p$ are the regression coefficient parameters for the p study characteristics, and u_i in Equation 9.2 is the deviation between the true effect size for ith study and the effect size that is predicted based on the study characteristics that are included in the regression model. Each random effect, u_i, is assumed to be independent with mean zero and variance σ_θ^2.

If the fixed-effects model is the model of choice, Equation 9.1 can be estimated using

$$\hat{\theta}_i = \beta_0 + \beta_1 X_{i1} + \ldots + \beta_p X_{ip} + \varepsilon_i, \tag{9.3}$$

where the estimation errors given by

$$\varepsilon_i = \hat{\theta}_i - \theta_i = \hat{\theta}_i - \left(\beta_0 + \beta_1 X_{i1} + \cdots + \beta_p X_{ip}\right). \tag{9.4}$$

The estimation errors ε_i are assumed to be statistically independent, each with mean zero and estimated variance $\text{Var}(\hat{\theta}_i)$.

If the random-effects model is the model of choice, Equation 9.2 can be estimated using

$$\hat{\theta}_i = \beta_0 + \beta_1 X_{i1} + \ldots + \beta_p X_{ip} + \varepsilon_i + u_i, \tag{9.5}$$

where the estimation errors given by

$$\varepsilon_i + u_i = \hat{\theta}_i - \theta_i = \hat{\theta}_i - \left(\beta_0 + \beta_1 X_{1i} + \cdots + \beta_p X_{pi} \right). \tag{9.6}$$

The estimation errors $\varepsilon_i + u_i$ that are assumed to be statistically independent, each with mean zero and estimated variance $\mathrm{Var}\left(\hat{\theta}_i\right) + \hat{\sigma}_\theta^2$, where $\hat{\sigma}_\theta^2$ is the estimator of $\hat{\sigma}_\theta^2$.

To help determine which model to choose, you can test the null hypothesis that random-effects variance σ_θ^2 equals zero. We assume that you have adequate power to conduct the homogeneity test. If you reject the null hypothesis that σ_θ^2 equals zero, the inference is that significant variation in the random effects remains after controlling for study characteristics. That is, you should use a random-effects model. If you fail to reject the null hypothesis that σ_θ^2 equals zero, the inference is that the study characteristics account for the variation in the random effects. That is, you should choose a fixed-effects model unless you want to make generalizations to a diverse group of studies. In this chapter, we describe how to test the null hypothesis that σ_θ^2 equals zero for three types of random-effects models: (a) one-factor ANOVA, (b) two-factor ANOVA, and (c) regression.

If you reject the null hypothesis that the random-effects variance σ_θ^2 equals zero, it is desirable to obtain an estimate of σ_θ^2. In this book, we use the SAS MIXED procedure to estimate σ_θ^2 (SAS, 1997). PROC MIXED implements two likelihood-based methods: maximum likelihood (ML) and residual (restricted) maximum likelihood (REML) (see McLachlan and Krishnan, 1997, for a detailed discussion of these two methods). A favorable theoretical property of both methods is that they accommodate data that are missing at random (Rubin 1976; Little, 1995). Because of its desirable statistical properties, the REML method is the default in PROC MIXED. The REML method is also the one that we use in this book to estimate the random-effects variance σ_θ^2.

The computation procedures that are provided in this chapter for random-effects models are very similar to those provided in the previous chapter for fixed-effects models, except that the weights differ. In the random-effects model, the effect-size estimate in the ith study is weighted by $1/\{\mathrm{Var}(\hat{\theta}_i)+\hat{\sigma}^2_\theta\}$. In a fixed-effects model, the effect-size estimate in the ith study is weighted by $1/\mathrm{Var}(\hat{\theta}_i)$ (see Chapter 8, Section 8.4.1.1).

9.2 Testing the Moderating Effects of Categorical Study Characteristics in ANOVA Models

This section focuses on testing the moderating effects of categorical study characteristics in random-effects ANOVA models. We consider random-effects ANOVA models with one and two categorical factors. Section 9.3 focuses on testing the moderating effects of continuous study characteristics in random-effects regression models.

9.2.2 Random-Effects Models with One Categorical Factor

Suppose that you are interested in testing the moderating effect of one categorical study characteristic with p independent groups, with m_1 effects in group 1, m_2 effects in group 2, ..., m_p effects in group p, and $k = m_1 + m_2 + \cdots + m_p$. The only difference in the computation procedures between the fixed-effects one-factor ANOVA model described in Chapter 8, Section 8.4.1, and the random-effects one-factor ANOVA models that are described in this section is the weight that is used in the analysis (see Section 9.1 for a description of how the effect-size estimate weights differ in random- and fixed-effects models).

9.2.2.1 Testing Whether the Random-Effects Variance is Zero

You can test the following null hypothesis that the random-effects variance is zero,

$$H_0 : \sigma_\theta^2 = 0, \tag{9.7}$$

against the alternative hypothesis that it is greater than zero,

$$H_A : \sigma_\theta^2 > 0. \tag{9.8}$$

The appropriate test statistic for Equation 9.7 is Q_{WITHIN} that was used for testing the heterogeneity of effect sizes after controlling for a single factor (see Chapter 8, Section 8.4.1.1). You reject the null hypothesis in Equation 9.7 at significance level α if Q_{WITHIN} exceeds the $100(1-\alpha)\%$ critical value of the chi-square distribution with $(k-p)$ degrees of freedom, where k is the number of independent studies and p is the number of independent groups of the categorical study characteristic.

Example 9.1 *Testing whether the random-effects variance is zero in a one-factor model*

The data for this example were obtained from a meta-analysis by Bettencourt and Miller (1996) on sex differences in aggression as a function of provocation, the same data set used for Example 8.1. This example includes two factors: aggression measure (that is, physical, evaluation, money loss) and provocation (that is, provoked versus neutral). In the SAS data set for this example, the aggression measure variable is called AGGR and the provocation variable is called PROV. In this section, AGGR is used as the single factor in the model. In Section 9.2.3, both factors are included in the model.

You can access the data set for this example, CH8.EX81, via SAS Online Samples (see "Using This Book"). The following SAS code was used to test the null hypothesis in Equation 9.1 that the random-effects variance is zero. The results are printed in Output 9.1.

```
options nodate nocenter pagesize=54 linesize=80 pageno=1;
libname ch8 "d:\metabook\ch8\dataset";
libname ch9 "d:\metabook\ch9\dataset";
data temp;
   set ch8.ex81;
   weight = 1 /veff;
proc glm data=temp noprint outstat=ch9.ex91out;
   class aggr;
   model eff=aggr;
   weight weight;
data ch9.ex91out;
   source="Error";
   set ch9.ex91out;
   if (_type_ = "ERROR");
   qstat=ss;
   pvalue=1-probchi(qstat,df);
   keep source qstat df pvalue;
proc print data=ch9.ex91out noobs;
   title;
   format qstat pvalue 8.4;
run;
```

Output 9.1 *Testing whether the random-effects variance is zero*

SOURCE	DF	QSTAT	PVALUE
Error	61	128.3896	0.0000

As you can see in Output 9.1, the null hypothesis that the random-effects variance is zero should be rejected at the .05 significance level because the *p*-value associated with the Q_{WITHIN} test is less than .05. Thus, a random-effects one-factor ANOVA model is more appropriate than a fixed-effects one-factor ANOVA model for this data set. Note that Q_{WITHIN} in this example is equal to the within-groups sum of squares in Output 8.1.

9.2.2.2 Estimating the Random-Effects Variance

We use the SAS MIXED procedure to estimate the random-effects variance σ_θ^2. Example 9.2 illustrates how to estimate the random-effects model variance using the residual (restricted) maximum likelihood (REML) method.

Example 9.2 *Estimating the random-effects model variance using the method of residual (restricted) maximum likelihood*

The following SAS code was used to estimate the variance for the random-effects one-factor ANOVA model in Example 9.1. The SAS code needed to obtain an estimate of the random-effects variance is given below. We only describe the SAS code that you need to change for your own data set. The variable names for the effect-size estimate and its estimated variance are EFF and VEFF, respectively. You need to replace `aggr` in the CLASS and MODEL statements with the variable name of the single categorical factor that is being tested. You also need to change the names of the SAS data sets in lines 7, 25, 26, and 31. The input data set CH8.EX81 is explained in Chapter 8 (see Example 8.1). The output SAS data set CH9.EX92OUT contains four variables: ESTIMATE, LEVEL, LOWER, and UPPER. ESTIMATE is the REML estimate for the random-variance component, LEVEL is the significant level of the confidence interval (for example, 0.95 for a 95% confidence interval), LOWER is the lower bound of the confidence interval, and UPPER is the upper bound of the confidence interval. Output 9.2 shows the results from the REML method.

```
options nodate nocenter pagesize=54 linesize=80 pageno=1;
libname ch8 "d:\research\metabook\ch8\dataset";
libname ch9 "d:\research\metabook\ch9\dataset";
/* the following data steps are used to create a dummy   */
/* random effect variable call "OBSNUM"                  */
data temp;
    set ch8.ex81;
    obsnum=_n_;
/* The following data steps are used to create a         */
/* diagonal covariance matrix with the estimated         */
/* variance as the diagonal elements                     */
data gdatain;
    set temp;
    col=_n_;
    row=_n_;
    value=veff;
    keep col row value;
/* Estimating the random effect variance                 */
proc mixed data=temp alpha=0.05 method=reml cl;
    class aggr obsnum;
    model eff = aggr;
    random obsnum/gdata=gdatain;
    make "covparms" out=ch9.ex92out;
```

```
/* Producing the OUTPUT                                    */
data ch9.ex92out;
   set ch9.ex92out;
   method="REML";
   estimate=est;
   level=1-alpha;
   keep method level upper lower estimate;
proc print data=ch9.ex92out noobs;
   title;
   format level 4.2 estimate upper lower 6.3;
run;
```

Output 9.2 *Estimating the random-effects variance using the method of residual (restricted) maximum likelihood*

METHOD	ESTIMATE	LEVEL	LOWER	UPPER
REML	0.136	0.95	0.096	0.200

As you can see in Output 9.2, the residual (restricted) maximum likelihood estimate of the random-effects variance is 0.136 with a 95% confidence interval [0.098 to 0.200]. Note that the confidence interval does not include the value zero.

9.2.2.3 Tests of Heterogeneity

In a random-effects model, you also can perform heterogeneity tests similar to the ones used in fixed-effects models (see Chapter 8, Section 8.4.1.1), except that the weight differs.

Example 9.3 *Heterogeneity tests for a one-factor model*

The data for this example were obtained from a meta-analysis by Bettencourt and Miller (1996) on sex differences in aggression as a function of provocation. You can use the SAS macro in Appendix 8.1 to produce the heterogeneity summary statistics that are given in Output 9.4. Only lines 5, 6, 7, 10, and 18 of the SAS code need to be modified. Lines 5 and 6 were described in Example 9.2. In line 7, we change `veff` to `veff + estimate`, where `veff` is the name for $\text{Var}\!\left(\hat{\theta}_i\right)$ and `estimate` is the name for the random-effects variance estimate $\hat{\sigma}_\theta^2$ based on the REML method. Lines 10 and 18 specify the name of the SAS output data set CH9.EX93OUTA.

```
options nodate nocenter pagesize=54 linesize=80 pageno=1;
libname ch8 "d:\research\metabook\ch8\dataset";
libname ch9 "d:\research\metabook\ch9\dataset";
data temp;
    set ch8.ex81;
    if (_n_ =1) then set ch9.ex92out;
    veff = veff + estimate;
%let vlist1 = aggr;
%let vlist2 = aggr;
%within(temp,ch9.ex93outa,eff,veff,aggr,3);
proc format;
    value aa  1 = "Between Groups       "
              2 = "Within Groups        "
              3 = "  Within (Physical)  "
              4 = "  Within (Evaluation)"
              5 = "  Within (Money Loss)"
              6 = "Corrected Total      ";
proc print data=ch9.ex93outa noobs;
    title;
    format source aa. qstat pvalue 7.3;
run;
```

Output 9.3 *ANOVA table for heterogeneity summary statistics for Example 9.1*

SOURCE	DF	QSTAT	PVALUE
Between Groups	2	5.292	0.071
Within Groups	61	61.002	0.476
Within (Physical)	33	30.999	0.567
Within (Evaluation)	25	28.825	0.271
Within (Money Loss)	3	1.177	0.759
Corrected Total	63	66.293	0.364

As you can see in Output 9.3, the between-groups heterogeneity test is nonsignificant, indicating that the magnitude of effect-size estimates does not differ as a function of the measure of aggression used. The within-groups heterogeneity test is also nonsignificant, indicating that the effects are homogenous.

The SAS macro in Appendix 8.2 can compute the effect-size estimates and confidence intervals for random-effects ANOVA models with one categorical factor. Only lines 5, 6, 12, 13, 14, and 21 of the SAS code need modification. The modifications made in both lines are similar to the modifications that are made in the SAS code for Output 9.3. Output 9.4 shows the results from the following SAS code:

```
options nodate nocenter pagesize=54 linesize=80 pageno=1;
libname ch8 "d:\metabook\ch8\dataset";
libname ch9 "d:\metabook\ch9\dataset";
data temp;
    set ch8.ex81;
    if (_n_ = 1) then set ch9.ex92out;
    veff = veff + estimate;
    output;
    aggr = 4;
    output;
%let varlist = aggr;
%wavgeff(temp,ch9.ex93outb,.95);
data ch9.ex93outb;
    set ch9.ex93outb;
    rename fact1=aggr;
proc format;
    value aa 1 = "Physical  "
             2 = "Evaluation"
             3 = "Money Loss"
             4 = "Over All  ";
proc print data=ch9.ex93outb noobs;
    var aggr estimate level lower upper;
    title;
    format aggr aa. estimate level lower upper 6.3;
run;
```

Output 9.4 Sex difference 95% confidence intervals for different measures of aggression and for all measures combined

AGGR	ESTIMATE	LEVEL	LOWER	UPPER
Physical	0.304	0.950	0.119	0.490
Evaluation	0.166	0.950	-0.032	0.365
Money Loss	-0.267	0.950	-0.726	0.192
Over All	0.199	0.950	0.069	0.329

Because the "between-groups" and "within-groups" heterogeneity tests were both nonsignificant (see Output 9.3), you should use the overall measure of aggression in Output 9.4. As you can see in Output 9.4, there was a small sex difference in overall aggression, with higher levels of aggression among males.

9.2.2.4 Comparisons among Group Mean Effects

If the factor in a random-effects model contains more than two levels, the omnibus $Q_{BETWEEN}$ statistic indicates that the weighted group mean effect sizes are not all equal. You can conduct *a priori* (planned) or *post hoc* (unplanned) contrasts to determine how the group weighted mean effect sizes differ. You should conduct *post hoc* comparisons only if the $Q_{BETWEEN}$ statistic is significant. You can conduct *a priori* contrasts regardless of whether the $Q_{BETWEEN}$ statistic is significant.

9.2.2.4.1 A priori contrasts among group mean effects. You can perform *a priori*, or planned, contrasts among group mean effects using the same procedures that are described in Chapter 8, Section 8.4.1.2.1, but the weights differ.

Example 9.4 *A priori contrasts among group mean effects.*

A planned contrast was conducted to compare the physical and evaluation aggression measures with the money loss aggression measure. The coefficients for this contrast are $c_1 = c_2 = 0.5$ and $c_3 = -1$. The following SAS code produces the results in Output 9.5:

```
options nodate nocenter pagesize=54 linesize=80 pageno=1;
libname ch9 "d:\research\metabook\ch9\dataset";
data tmpin;
    set ch9.ex93outb;
    if (aggr ^= 4);
%let varlist = estimate variance;
data coeff;
    input c1 c2 c3;
cards;
0.5    0.5    -1
;
%fixconst(tmpin,tmpout,coeff,3,0.95)
proc format;
    value aa 1 = "Planned Comparison";
proc print data=tmpout noobs;
    format type aa. contrast level lower upper 7.3;
quit;
```

Output 9.5 *Planned comparison using the average of physical and evaluation versus money loss*

TYPE	CONTRAST	LEVEL	LOWER	UPPER
Planned Comparison	0.502	0.950	0.023	0.981

As you can see in Output 9.5, the contrast was significant at the .05 level because the confidence interval does not include the value zero. Sex differences in physical and evaluation measures of aggression were different from sex differences in the money-loss measure of aggression. You can see the nature of these sex differences in Output 9.5.

9.2.2.4.2 Post hoc comparisons among group mean effects. You can perform *post hoc* or unplanned comparisons among group mean effects using the same procedures that are described in Chapter 8, Section 8.4.1.2.2, but the weights differ. We do not illustrate how to conduct *post hoc* comparisons because the between-groups Q-statistic was nonsignificant in Output 9.3.

9.2.3 Random-Effects Models with Two Categorical Factors

Random-effects two-factor ANOVA models are similar to fixed-effects two-factor ANOVA models, except the weights differ.

9.2.3.1 Testing Whether the Random-Effects Variance is Zero

You can test the following null hypothesis that the random-effects variance is zero,

$$H_0 : \sigma_\theta^2 = 0, \tag{9.9}$$

against the alternative hypothesis that it is greater than zero,

$$H_A : \sigma_\theta^2 > 0. \tag{9.10}$$

The appropriate test statistic for Equation 9.9 is Q_{WITHIN} that was used for testing the heterogeneity of effect sizes after controlling for two factors (see Chapter 8, Section 8.4.2). You reject the null hypothesis in Equation 9.9 at significance level α if Q_{WITHIN} exceeds the $100(1-\alpha)\%$ critical value of the chi-square distribution with $(k - a \times b)$ degrees of freedom, where k is the number of independent studies, a is the number of groups in Factor A, and b is the number of groups in Factor B.

Example 9.5 *Testing whether the random-effects variance is zero in a two-factor mode.*

The data for this example were obtained from the meta-analysis by Bettencourt and Miller (1996) on sex differences in aggression as a function of provocation. The two factors included in the model were aggression measure (that is, physical, evaluation, money loss) and provocation (that is, provoked versus neutral). The following SAS code was used to produce the results in Output 9.6. Note that the provocation variable PROV was added to the CLASS statement. The MODEL statement also contains the main effect for provocation (that is, PROV) and the interaction between aggression measure and provocation (that is, AGGR*PROV).

```
libname ch8 "d:\metabook\ch8\dataset";
libname ch9 "d:\metabook\ch9\dataset";
data;
    set ch8.ex81;
    weight = 1 / veff;
proc glm noprint outstat=ch9.ex95out;
    class aggr prov;
    model eff=aggr prov aggr*prov;
    weight weight;
data ch9.ex95out;
    source="Error";
    set ch9.ex95out;
    if (_type_ = "ERROR");
    qstat=ss;
    pvalue=1-probchi(qstat,df);
    keep source qstat df pvalue;
proc print data=ch9.ex95out noobs;
    title;
    format qstat pvalue 8.4;
run;
```

Output 9.6 *Testing whether the random-effects variance is zero*

SOURCE	DF	QSTAT	PVALUE
Error	58	123.2023	0.0000

As you can see in Output 9.6, the null hypothesis that the random-effects variance is zero should be rejected at the .05 significance level because the p-value associated with the Q_{WITHIN} test is less than .05. Thus, a random-effects two-factor ANOVA model is more appropriate than a fixed-effects two-factor ANOVA model for this data set.

9.2.3.2 Estimating the Random-Effects Variance

Example 9.6 illustrates how to estimate the random-effects model variance using the REML method.

Example 9.6 *Estimating the random-effects model variance using the method of residual (restricted) maximum likelihood*

The SAS code for this example is the same as the SAS code for Example 9.2 except that the CLASS statement now contains the provocation variable (that is, PROV) and the MODEL statement now contains the main effect for provocation (that is, PROV) and the interaction between aggression measure and provocation (that is, AGGR*PROV). Output 9.7 shows the results from the following SAS code:

```
options nodate nocenter pagesize=54 linesize=80 pageno=1;
libname ch8 "d:\research\metabook\ch8\dataset";
libname ch9 "d:\research\metabook\ch9\dataset";
data temp;
   set ch8.ex81;
   obsnum=_n_;
```

```
data gdatain;
   set temp;
   col=_n_;
   row=_n_;
   value=veff;
   keep col row value;
proc mixed data=temp alpha=0.05 method=reml cl;
   class aggr prov obsnum;
   model eff = aggr prov;
   random obsnum/gdata=gdatain;
   make "covparms" out=ch9.ex96out;
data ch9.ex96out;
   set ch9.ex96out;
   method="REML";
   estimate=est;
   level=1-alpha;
   keep method level upper lower estimate;
proc print data=ch9.ex96out noobs;
   title;
   format level 4.2 estimate upper lower 6.3;
run;
```

Output 9.7 *Estimating the random-effects variance using the method of residual (restricted) maximum likelihood*

METHOD	ESTIMATE	LEVEL	LOWER	UPPER
REML	0.127	0.95	0.092	0.189

As you can see in Output 9.7, the residual (restricted) maximum likelihood estimate of the random-effects variance in the two-factor ANOVA model is 0.127 with 95% confidence interval [0.092, 0.189]. Note that the confidence interval does not include the value zero.

9.2.3.3 Heterogeneity Tests

In the two-factor random-effects ANOVA, the total heterogeneity statistic can be partitioned into four components: (a) heterogeneity between levels of factor A (that is, Q_A), (b) heterogeneity between levels of factor B (that is, Q_B), (c) heterogeneity due to the interaction AB (that is, Q_{AB}), and (d) heterogeneity within groups of studies (that is, Q_{WITHIN}). The computations are the same as for a two-factor fixed-effects ANOVA (see Chapter 8, Section 8.4.2), except the weights differ (see Section 9.1 for a discussion of how the weights differ in random- and fixed-effects models).

Example 9.7 *Tests of heterogeneity*

The following SAS code produces the heterogeneity tests for a two-factor random-effects ANOVA model. Output 9.8 shows the results.

```
libname ch8 "d:\metabook\ch8\dataset";
libname ch9 "d:\metabook\ch9\dataset";
data temp;
    set ch8.ex81;
    if (_n_ = 1) then set ch9.ex96out;
    veff = veff + estimate;
    if (aggr = 1) then trt=aggr+prov-1;
    else if (aggr = 2) then trt=aggr+prov;
    else if (aggr = 3) then trt=aggr+prov+1;
%let vlist1 = aggr prov;
%let vlist2 = aggr prov aggr*prov;
%within(temp,ch9.ex97out,eff,veff,trt,6);
proc format;
    value aa    1 = "Aggression                        "
                2 = "Provocation                       "
                3 = "Aggression*Provocation            "
                4 = "Within                            "
                5 = "  Physical and Provoked           "
                6 = "  Physical and Neutral            "
                7 = "  Evaluation and Provoked         "
                8 = "  Evaluation and Neutral          "
                9 = "  Money Loss and Provoked         "
               10 = "  Money Loss and Neutral          "
               11 = "Corrected Total                   ";
proc print data=ch9.ex97out noobs;
    title;
    format source aa. qstat pvalue 7.3;
quit;
```

Output 9.8 *Heterogeneity tests for a two-factor ANOVA model.*

```
SOURCE                          DF      QSTAT      PVALUE
Aggression                       2       5.476      0.065
Provocation                      1       2.612      0.106
Aggression*Provocation           2       0.889      0.641
Within                          58      59.111      0.435
  Physical and Provoked         16      12.425      0.714
  Physical and Neutral          16      19.173      0.260
  Evaluation and Provoked       12      12.806      0.383
  Evaluation and Neutral        12      14.316      0.281
  Money Loss and Provoked        1       0.377      0.539
  Money Loss and Neutral         1       0.015      0.902
Corrected Total                 63      68.088      0.308
```

As you can see in Output 9.8, none of the heterogeneity tests are significant.

9.3 Testing the Moderating Effects of Continuous Study Characteristics in Regression Models

This section focuses on testing the moderating effects of continuous study characteristics in random-effects regression models. The only computational difference between the fixed-effects regression models that are described in Chapter 8, Section 8.5, and the random-effects regression models that are described in this section is that the effect-size estimates are weighted by $1/\{\mathrm{Var}(\hat{\theta}_i) + \hat{\sigma}_\theta^2\}$ rather than by $1/\mathrm{Var}(\hat{\theta}_i)$.

9.3.1 Testing Whether the Random-Effects Variance is Zero

You can test the following null hypothesis that the random-effects variance is zero,

$$H_0 : \sigma_\theta^2 = 0, \tag{9.11}$$

against the alternative hypothesis that it is greater than zero,

$$H_A : \sigma_\theta^2 > 0. \tag{9.12}$$

The appropriate test statistic for Equation 9.11 is Q_{ERROR} that was used for testing the heterogeneity of effect sizes after controlling for the p study characteristics (see Chapter 8, Section 8.5.2). You reject the null hypothesis in Equation 9.11 at significance level α if Q_{ERROR} exceeds the $100(1-\alpha)\%$ critical value of the chi-square distribution with $(k-p-1)$ degrees of freedom.

Example 9.8 *Testing whether the random effects variance is zero in the meta-analysis on group cohesiveness and performance*

Example 9.8 uses the meta-analysis on the relation between group cohesiveness and performance (Mullen & Copper, 1994), the same data set used for Example 8.4. Recall that Mullen and Cooper wanted to test the hypothesis that cohesive groups perform better on tasks than do noncohesive groups. They coded four study characteristics: group size (continuous), whether the group was real (coded 1) or artificial (coded 0), whether the group required high (coded 1) or low (coded 0) interaction, and whether the study was correlational (coded 1) or experimental (coded 0).

You can access the data set for this example, CH8.EX84, via SAS Online Samples (see "Using This Book.") The following SAS code was used to test the null hypothesis in Equation 9.7 that the random effects variance is zero. Output 9.9 shows the results.

```
options nodate nocenter pagesize=54 linesize=80 pageno=1;
libname ch8 "d:\research\metabook\ch8\dataset";
libname ch9 "d:\research\metabook\ch9\dataset";
data temp;
  set ch8.ex84;
    weight = 1 / veff;
proc glm data=temp noprint outstat=ch9.ex98out;
    weight weight;
    model eff = group reality paradigm interact;
data ch9.ex98out;
    source = "Error";
    set ch9.ex98out;
    if (_type_ = "ERROR");
    qstat=ss;
    pvalue=1-probchi(qstat,df);
    keep source df qstat pvalue;
```

```
proc print data=ch9.ex98out noobs;
   title;
   format qstat pvalue 8.4;
quit;
```

Output 9.9 *Testing whether the random effects variance is zero*

SOURCE	DF	QSTAT	PVALUE
Error	61	420.1070	0.0000

As you can see in Output 9.9, the null hypothesis that the random-effects variance is zero should be rejected at the .05 significance level because the p-value associated with the Q_{ERROR} test is less than .05. Thus, a random-effects regression model is more appropriate than a fixed-effects regression model for this data set.

9.3.2 Estimating the Random-Effects Variance

The random-effects variance in regression models can also be estimated using the method of residual (restricted) maximum likelihood.

Example 9.9 *Estimating the random-effects variance*

The following SAS code was used to estimate the random-effects variance in Example 9.8 using the method of residual (restricted) maximum likelihood. This SAS code is similar to previous examples. Output 9.10 shows the results.

```
options nodate nocenter pagesize=54 linesize=80 pageno=1;
libname ch8 "d:\research\metabook\ch8\dataset";
libname ch9 "d:\research\metabook\ch9\dataset";
data temp;
    set ch8.ex84;
    obsnum=_n_;
data gdatain;
    set temp;
    col=_n_;
    row=_n_;
    value=veff;
    keep col row value;
proc mixed data=temp alpha=0.05 method=reml cl;
    class group reality paradigm interact obsnum;
    model eff = group reality paradigm interact;
    random obsnum/gdata=gdatain;
    make "covparms" out=ch9.ex99out;
data ch9.ex99out;
    set ch9.ex99out;
    method="REML";
    estimate=est;
    level=1-alpha;
    keep method level upper lower estimate;
proc print data=ch9.ex99out noobs;
    title;
    format level 4.2 estimate upper lower 6.3;
run;
```

Output 9.10 *Estimating the random-effects variance using the method of residual (restricted) maximum likelihood*

METHOD	ESTIMATE	LEVEL	LOWER	UPPER
REML	0.045	0.95	0.030	0.075

As you can see in Output 9.10, the method of residual (restricted) maximum likelihood estimate of the random-effects variance is 0.045 with 95% confidence interval [0.030, 0.075]. Note that the confidence interval does not include the value zero.

9.3.3 Confidence Intervals for Individual Regression Coefficients

The SAS code that is used to compute confidence intervals and hypothesis tests for random-effects models is similar to the SAS code that is used for fixed-effects models, except the weights differ.

Example 9.10 *Confidence intervals for individual regression coefficients in the meta-analysis on the relation between group cohesiveness and performance*

The SAS macro in Appendix 8.4, along with the following SAS code, was used to obtain confidence intervals for individual regression coefficients. The only modification made to the SAS code was the addition of lines 6 and 7. Output 9.11 shows the results.

```
options nodate nocenter pagesize=54 linesize=80 pageno=1;
libname ch9 "d:\metabook\ch9\dataset";
libname ch8 "d:\metabook\ch8\dataset";
data temp;
   set ch8.ex84;
   if (_n_ = 1) then set ch9.ex99out;
   veff = veff + estimate;
   keep group reality paradigm interact eff veff;
%fixreg(temp,ch9.ex910out,.95,eff,veff);
proc format;
   value aa 1 = "Intercept"
            2 = "Group    "
            3 = "Reality  "
            4 = "Paradigm "
            5 = "Interact ";
proc print data=ch9.ex910out noobs;
   title;
   format param aa. estimate lower upper 6.3;
run;
```

Output 9.11 *Confidence intervals for individual regression coefficients in the meta-analysis on group cohesiveness and performance*

PARAM	ESTIMATE	LEVEL	LOWER	UPPER
Intercept	0.303	0.95	0.131	0.474
Group	-0.026	0.95	-0.048	-0.004
Reality	0.274	0.95	-0.048	-0.004
Paradigm	-0.050	0.95	-0.230	0.130
Interact	0.040	0.95	-0.0115	0.195

As you can see in Output 9.11, the confidence intervals for group size and group realism do not include the value zero. As group size increased, the relation between group cohesiveness and performance decreased. The relation between group cohesiveness and performance was stronger for real groups than for artificial groups. The confidence intervals for PARADIGM (that is, correlational study versus experimental study) and INTERACT (that is, high versus low interaction) both included the value zero.

9.3.4 Omnibus Tests for Blocks of Regression Coefficients

The omnibus tests for blocks of regression coefficients are computed the same way as in Chapter 8, Section 8.5.2, except the weights differ.

Example 9.11 Omnibus tests for blocks of regression coefficients

Because the individual regression coefficients for the study characteristics PARADIGM and INTERACT were nonsignificant (see Example 9.10), an analysis was conducted to determine whether these study characteristics should be excluded from the regression model. In this analysis, GROUP and REAL were entered as the first block of variables, and PARADIGM and INTERACT were entered as the second block of variables. The following SAS code was used to produce the results in Output 9.12 with the random-effects variance estimated using the method of residual (restricted) maximum likelihood. Note that the weight statement was changed in line 7 of the SAS code.

```
options nodate nocenter pagesize=54 linesize=80 pageno=1;
libname ch8 "d:\metabook\ch8\dataset";
libname ch9 "d:\metabook\ch9\dataset";
data temp;
   set ch8.ex84;
   if (_n_ = 1) then set ch9.ex99out2;
   weight = 1 / (veff+estimate);
proc glm data=temp noprint outstat=aa;
   weight weight;
   model eff = reality group paradigm interact;
data cc dd;
   set aa;
   qstat=ss;
   if (_type_ = "SS1  " and (_source_  = "PARADIGM"
      or _source_ = "INTERACT")) then output cc;
   if ((_type_ = "SS1  " and (_source_  = "PARADIGM" or
      _source_ = "INTERACT")) or _type_ = "ERROR") then
output dd;
   keep df qstat;
proc means data=cc noprint;
   var df qstat;
   output out=aa sum=df qstat;
proc means data=dd noprint;
   var df qstat;
   output out=bb sum=df qstat;
data data=ch9.ex911out;
   source=_n_;
   set aa bb;
   pvalue=1-probchi(qstat,df);
   keep source df qstat pvalue;
proc format;
   value aa 1 = "QCHANGE"
            2 = "QERROR ";
proc print data=ch9.ex911out noobs;
   title;
   format source aa. qstat pvalue 7.3;
run;
```

Output 9.12 Omnibus tests for blocks of regression coefficients in the meta-analysis on group cohesiveness and performance

SOURCE	DF	QSTAT	PVALUE
QCHANGE	2	0.463	0.793
QERROR	63	67.770	0.318

Because the statistic Q_{CHANGE} in Output 9.12 is nonsignificant at the .05 level, we fail to reject the null hypothesis that the PARADIGM and INTERACT study characteristics are not related to effect sizes after controlling for the GROUP and REALITY study characteristics. Thus, you probably should exclude the study characteristics PARADIGM and INTERACT from the regression model. The Q_{ERROR} statistic is also nonsignificant, suggesting that the effect-size estimates are homogenous and can, therefore, be combined.

Finally, we perform a regression analysis that only includes the two study characteristics with significant regression coefficients (that is, group size and whether the group was real or artificial). Output 9.13 shows the results of the following SAS code:

```
options nodate nocenter pagesize=54 linesize=80 pageno=1;
libname ch9 "d:\research\metabook\ch9\dataset";
libname ch8 "d:\research\metabook\ch8\dataset";
data temp;
    set ch8.ex84;
    if (_n_ = 1) then set ch9.ex99out;
    veff = veff + estimate;
    keep group reality eff veff;
%fixreg(temp,ch9.ex912out,.95,eff,veff);
proc format;
    value aa 1 = "Intercept"
             2 = "Group    "
             3 = "Reality  ";
proc print data=ch9.ex912out noobs;
    title;
    format param aa. estimate lower upper 6.3;
run;
```

Output 9.13 *Confidence intervals for individual regression coefficients*

PARAM	ESTIMATE	LEVEL	LOWER	UPPER
Intercept	0.318	0.95	0.195	0.441
Group	-0.026	0.95	-0.048	-0.004
Reality	0.241	0.95	0.099	0.382

As you can see in Output 9.13, the confidence intervals for group size and group realism do not include the value zero. As group size increased, the relation between group cohesiveness and performance decreased. The relation between group cohesiveness and performance was stronger for real groups than for artificial groups. This reduced model seems to provide a good fit to the data because the Q_{ERROR} statistic is nonsignificant (see Output 9.12).

9.3.5 Multicollinearity among Study Characteristics

Multicollinearity can be tested by means of variance inflation factors, VIF (see Neter, Wasserman, & Kutner, 1990). A VIF of 1 indicates that the study characteristics are not linearly related. A maximum VIF value in excess of 10 is often taken as an indication that multicollinearity may be unduly influencing the least squares estimate. You can compute a VIF for each study characteristic in the random-effects regression model using Equation 8.42, except with different weights.

Example 9.12: *Testing for multicollinearity among study characteristics in the meta-analysis on group cohesiveness and performance.*

The following SAS code was used to obtain the VIF for each study characteristic in the meta-analysis on group cohesion and performance (Mullen & Cooper, 1994) in a random-effects regression model with the variance estimated using the method of residual (restricted) maximum likelihood. Output 9.14 shows the results.

```
options nodate nocenter pagesize=54 linesize=80 pageno=1;
libname ch8 "d:\metabook\ch8\dataset";
libname ch9 "d:\metabook\ch9\dataset";
data temp;
    set ch8.ex84;
    if (_n_ = 1) then set ch9.ex99out2;
    weight = 1 / (veff+estimate);
proc glm data=temp noprint outstat=aa;
    weight weight;
    model eff = reality group paradigm interact;
data temp;
    set aa;
    if (_type_ ^= "SS3  ");
proc means data=temp noprint;
    var ss;
    output out=bb sum=ssct;
data ch9.ex912;
    set aa;
    source=_n_-5;
    if (_type_ = "SS3  ");
    keep source ss;
data temp;
    set ch9.ex912;
    if (_n_ = 1) then set bb;
    vif = 1 / (1-ss/ssct);
    keep source vif;
proc format;
    value aa 1 = "Reality "
             2 = "Group   "
             3 = "Paradigm"
             4 = "Interact";
proc print data=temp noobs;
    format source aa. vif 5.3;
    title;
quit;
```

Output 9.14 *Variance inflation factors*

```
SOURCE       VIF

Reality      1.131
Group        1.072
Paradigm     1.004
Interact     1.003
```

As you can see in Output 9.14, the maximum VIF is 1.131, a value only slightly larger than 1 and much smaller than 10. Thus, multicollinearity is not a problem in this example.

9.4 Conclusions

The random-effects model assumes that the population effect size is randomly distributed with its own mean and variance. In this chapter we used the method of residual (restricted) maximum likelihood in PROC MIXED to estimate the random-effects variance σ_θ^2. Cook et al. (1992) recommend using random-effects models when: (a) studies are heterogeneous, (b) treatments are ill-specified, and (c) treatment effects are complex and multi-determined. With random-effects models, you can make generalizations to a universe of such diverse studies. Although random-effects have greater generalizability than fixed-effects models, fixed-effects models have greater statistical power than random-effects models. If the variation in effects between studies can be explained by a few simple study characteristics, then a fixed-effects model might be more appropriate than a random-effects model (see Chapter 8 for a discussion of fixed-effects models in meta-analysis). We recommend that meta-analysts fit their data sets using both fixed- and random-effects models.

9.5 References

Bettencourt, B. A., & Miller, N. (1996). Gender differences in aggression as a function of provocation: A meta-analysis. *Psychological Bulletin, 119*, 422–447.

Cochran, W. G., & Cox, G. M. (1957). *Experimental designs* (2nd ed.). New York: Wiley.

Cook, T. D., Cooper, H., Cordray, D. S., Hartmann, H., Hedges, L. V., Light, R. J., Louis, T. A., & Mosteller, F. (1992). *Meta-analysis for explanation: A casebook.* New York: Russell Sage Foundation.

Graybill, F. G. (1976). *Theory and application of the linear model.* Pacific Grove, CA: Wadsworth & Brooks/Cole.

Hedges, L. V., & Olkin, I. (1985). *Statistical methods for meta-analysis.* New York: Academic Press.

Little, R. J. A. (1995). Modeling the drop out mechanism in repeated-measures studies. *Journal of the American Statistical Association, 83*, 1014–1022.

McLachlan, G. J., & Krishnan, T. (1997). *The EM algorithm and extensions.* New York: NY. John Wiley & Sons.

Mullen, B., & Copper, C. (1994). The relation between group cohesiveness and performance: An integration. *Psychological Bulletin, 115*, 210–227.

Neter, J., Wasserman, W., & Kutner, M. H. (1990). *Applied linear statistical models* (3rd ed.). Homewood, IL: Irwin.

Raudenbush, S. W. (1994). Random effects models. In H. Cooper & L. V. Hedges (Eds.). *The handbook of research synthesis* (pp. 301–321). New York: Russell Sage Foundation.

Rosenthal, R. (1995). Writing meta-analytic reviews. *Psychological Bulletin, 118*, 183–192.

Rubin, D. B. (1976). Inference and missing data. *Biometrika, 63*, 581–592.

SAS Institute Inc. (1996). *SAS/STAT software: Changes and enhancements* (through Release 6.11). Cary: SAS Institute Inc.

Searle, S. R. (1971). *Linear models.* New York: John Wiley & Sons.

chapter 10

Combining Correlated Effect-Size Estimates

10.1 Introduction

Most of the meta-analytic procedures that are described in this book are concerned with combining effect-size estimates obtained from k independent studies where each study compares a single treatment (experimental) group with a single control group. Because the k studies are statistically independent, so are their corresponding effect-size estimates. Sometimes, however, effect-size estimates are not independent. For example, consider two hypothetical studies designed to test the hypothesis that frustration increases aggression (Dollard, Doob, Miller, Mowrer, & Sears, 1939). In these studies, frustration is defined as preventing someone from obtaining a desired goal, and aggression is defined as behavior that is intended to harm someone.

Study 1

In the first hypothetical study, participants were randomly assigned to task frustration, personal frustration, or control groups. In the task frustration group, participants were given five minutes to complete an unsolvable jigsaw puzzle in the presence of a passive confederate (the researcher's accomplice). In the personal frustration group, participants were given five minutes to complete a solvable jigsaw puzzle, but a confederate prevented them from completing the puzzle in the allotted time. In the control group, participants were given five minutes to complete a solvable jigsaw puzzle in the presence of a passive confederate. Participants were then given an opportunity to shock the confederate for errors made on a different task. The shocks ranged in intensity from Level 1 (described as "just noticeable") to Level 10 (described as "definitely unpleasant"). The measure of aggression was the level of shock participants gave the confederate.

Study 2

In the second hypothetical study, a confederate was parked at a busy intersection. Just before the traffic light turned green, the confederate flipped a coin to determine whether to remain stalled at the intersection ten seconds after the light turned green (frustration condition) or to drive through the intersection the instant the light turned green (control condition). An observer in a nearby car recorded the frequency and duration of horn honks and the frequency of verbal comments (for example, cursing the confederate) and nonverbal gestures (for example, shaking a fist at the confederate in anger) from motorists who were behind the confederate.

Gleser and Olkin (1994) identified two types of studies that yield dependent or correlated effect-size estimates: multiple-treatment studies and multiple-endpoint studies. Multiple-treatment studies compare more than one treatment with a common control. An effect-size estimate is calculated for each treatment versus control comparison. Because of the common control group, the effect-size estimates will be correlated. In Study 1, for example, the researcher used two frustration groups (that is, task frustration, personal frustration) and a single control group.

In multiple-endpoint studies, there may be a single treatment group and a single control group, but multiple dependent variables are used as endpoints for each participant. A treatment versus control effect-size estimate is calculated for each dependent variable. Because measures on each participant are correlated, the effect-size estimates for the measures will be correlated within studies. In Study 2, for example, the researcher used four measures of aggression (that is, frequency of horn honks, duration of horn honks, frequency of verbal comments, frequency of nonverbal gestures). In the fields of education and psychology, one common type of multiple-endpoint study uses the correlated subscales of a test as the multiple endpoints. For example, the Scholastic Aptitude Test (SAT), a college entrance exam test, contains verbal and math subtests. Similarly, the Aggression Questionnaire (Buss & Perry, 1992) contains four correlated subscales that measure individual differences in aggressiveness (that is, physical aggression, verbal aggression, anger, and hostility).

The problem of dependent or correlated effect-size estimates has been dealt with in a number of ways (Kalaian & Raudenbush, 1996) . For example, in multiple-endpoint studies you could randomly select one of the effect-size estimates, select the most "representative" effect-size estimate, or use the average of all of the effect-size estimates. The disadvantage of this approach is that it willfully discards data – a wasteful practice. Another approach is to treat correlated effect-size estimates as if they were independent and then adjust the significance level. The disadvantage of this approach is that you do not know how large of an adjustment to make to the significance level. A Bonferroni adjustment, for example, might be too conservative. A better approach is to use multivariate procedures that take into account the correlated nature of the data.

In this chapter, we describe how to combine effect-size estimates from multiple-treatment and multiple-endpoint studies. We discuss each type of study in turn.

10.2 Combining the Results from Multiple-Treatment Studies

In multiple-treatment studies, independent groups of participants are randomly assigned to one of p treatment (experimental) groups or to a control group. Let the control group denoted by C and let the p experimental groups be denoted by $E_1,...,E_p$. You will rarely find a set of studies in which each study reports results for exactly the same set of experimental groups. It is more likely that different studies will report results for different subsets of experimental groups.

Table 10.1 gives an example of eight studies on alcohol-related aggression, each of which compared a control group to at least two experimental groups. Participants in the control group (that is, C) expected a nonalcoholic beverage and got a nonalcoholic beverage. Participants in the placebo group (that is, E_1) expected an alcoholic beverage but got a nonalcoholic beverage. Participants in the antiplacebo group (that is, E_2) expected a nonalcoholic beverage but got an alcoholic beverage. Participants in the alcohol group (that is, E_3) expected an alcoholic beverage and got an alcoholic beverage. An asterisk indicates whether a particular group was included in a study. For example, the study by Giancola and Zeichner (1995) included the control and experimental groups E_1 and E_3. Participants in all eight studies were provoked in some manner. In all eight studies, the aggression measure was the intensity of shock that participants gave the confederate who provoked them.

Table 10.1 *Studies on alcohol-related aggression that included a control group and at least two other experimental groups*

Study	C	E_1	E_2	E_3
Giancola & Zeichner, 1995	*	*		*
Gustafson, 1986	*	*	*	*
Gustafson, 1991	*	*		*
Lang et al., 1975	*	*	*	*
Pihl & Zacchia, 1986	*	*		*
Ratliff, 1984	*	*	*	*
Zeichner & Pihl, 1979	*	*		*
Zeichner & Pihl, 1980	*	*		*

Note: E_1 = placebo. E_2 = antiplacebo. E_3 = alcohol. * = group included in study.

For each participant in the control group, the dependent variable is assumed to have a normal distribution with mean μ_C and variance σ_C^2. For each participant in the jth experimental group, the dependent variable is assumed to have a normal distribution with mean μ_{E_j} and variance $\sigma_{E_j}^2$.

The population standardized mean difference for the jth treatment is

$$\delta_{E_j} = \frac{\mu_{E_j} - \mu_C}{\sigma_C}, \tag{10.1}$$

(Gleser & Olkin, 1994, p. 341). The sample estimator of δ_{E_j} is

$$d_{E_j} = \frac{\bar{Y}_{E_j} - \bar{Y}_C}{S_C}, \tag{10.2}$$

where $\bar{Y}_{E_j}$ is the sample mean for the jth experimental group, $\bar{Y}_C$ is the sample mean for the control group, and S_C is the sample standard deviation for the control group (Gleser & Olkin, 1994, p. 341). Note that the effect-size estimates $d_{E_1}, \ldots, d_{E_p}$ are correlated because they have $\bar{Y}_C$ and S_C in common.

When the control and experimental group variances are homogeneous within each study (that is, $\sigma_C^2 = \sigma_{E_1}^2 = \ldots = \sigma_{E_p}^2$), you can use the pooled standard deviation

$$S_{POOLED} = \sqrt{\frac{(n_C - 1)S_C^2 + \sum_{j=1}^{p}(n_{E_j} - 1)S_{E_j}^2}{(n_C - 1) + \sum_{j=1}^{p}(n_{E_j} - 1)}} \qquad (10.3)$$

instead of the control group standard deviation S_C in Equation 10.2.

Under the assumption of homogeneity of variance, the estimated large sample variance of d_{E_j} is

$$\mathrm{Var}\left(d_{E_j}\right) = \frac{1}{n_{E_j}} + \frac{1}{n_C} + \frac{\frac{1}{2}d_{E_j}^2}{N}, \qquad (10.4)$$

where $N = n_C + \sum_{j=1}^{p} n_{E_j}$ is the total sample size for the study in which the effect size is calculated (Gleser & Olkin, 1994, p. 346). If the variances are not homogeneous, the estimated large sample variance of d_{E_j} is

$$\mathrm{Var}\left(d_{E_j}\right) = \frac{1}{n_{E_j}}\left(\frac{S_{E_j}^2}{S_C^2}\right) + \frac{1 + \frac{1}{2}d_{E_j}^2}{n_C} \qquad (10.5)$$

(Gleser & Olkin, 1994, p. 346). Regardless of whether the homogeneity assumption is met, the estimated large sample covariance between d_{E_j} and $d_{E_{j^*}}$ $(j^* \neq j)$ is

$$\mathrm{Cov}\left(d_{E_j}, d_{E_{j^*}}\right) = \frac{1 + \frac{1}{2}d_{E_j}d_{E_{j^*}}}{n_C} \qquad (10.6)$$

(Gleser & Olkin, 1994, p. 341). To test the equal variance assumption, we use the statistic

$$F_{MAX} = \max_{j=1}^{p}\left(\frac{S_{E_j}^2}{S_C^2}\right), \qquad (10.7)$$

where $S_{E_j}^2$ and S_C^2 are the jth experimental group and control group variances for each study, respectively (Hartley, 1950). To compute combined effect-size estimates and 95% confidence intervals for each treatment, we use linear model theory (Gleser & Olkin, 1994; Graybill, 1976).

Example 10.1 *Multiple-treatment studies*

This example uses the same studies that are depicted in Table 10.1. The sample sizes, means, and standard deviations for each group included in a study are given in Table 10.2. The results in Table 10.2 were taken from previous meta-analyses on alcohol-related aggression (Bushman, 1993, 1997; Bushman & Cooper, 1990).

Table 10.2 *Descriptive statistics for studies on alcohol-related aggression that included a control group and at least two other experimental groups*

Study	Control			Placebo			Antiplacebo			Alcohol		
	N	M	SD	N	M	SD	N	M	SD	N	M	SD
Giancola & Zeichner, 1995	16	2.30	0.92	16	2.70	0.99				32	3.40	1.10
George & Marlatt, 1986	8	2.41	0.78	8	2.57	0.69	8	3.17	0.61	8	2.86	0.77
Gustafson, 1986	10	2.48	1.60	10	1.23	1.40	10	3.68	2.68	10	3.22	2.30
Gustafson, 1991	15	2.93	3.20	15	1.50	2.76				15	2.73	2.52
Lang et al., 1975	12	3.59	1.59	12	5.00	2.32	12	3.80	2.53	12	5.06	2.65
Pihl & Zacchia, 1986	16	2.48	0.58	16	2.78	0.79				16	2.37	0.74
Ratliff, 1984	6	2.40	0.94	6	2.70	1.04	6	2.60	1.06	6	3.00	0.92
Zeichner & Pihl, 1979	24	1.78	0.48	24	1.98	0.42				24	4.24	0.21
Zeichner & Pihl, 1980	24	1.68	0.26	24	2.01	0.41				24	4.19	0.16

Note: N = sample size, M = mean, SD = standard deviation.

First, we use SAS macro MTFMAX(*indata,outdata*) in Appendix 10.1 to test the equal variance assumption. *Indata* and *outdata* are the names of the input and output files, respectively. *Indata* contains the variable names for the experimental and control group standard deviations and sample sizes. This example uses three experimental groups (that is, 1 = placebo, 2 = antiplacebo, 3 = alcohol). The statement %let nelist = ne1 ne2 ne3 gives the variable names for the experimental group sample sizes. The statement %let nelist = ne1 ne2 ne3 gives the variable names for the experimental group standard deviations.

The statement %let nclist = nc gives the variable name for the control group sample size. The statement %let sc list = sc gives the variable name for the control group standard deviation.

Outdata contains five variables: (a) the means of identifying each study, either by number or by author names, STUDYID, (b) the F_{MAX} statistic, FMAX, (c) the F_{MAX} numerator degrees of freedom, NUMERDF, (d) the F_{MAX} denominator degrees of freedom, DENOMDF, and (e) the F_{MAX} p-value, PVALUE.

The following SAS code computes an F_{MAX} statistic for each study in Table 10.2. Output 10.1 shows the results.

```
libname ch10 "d:\metabook\ch10\dataset";
%let nelist = ne1 ne2 ne3;
%let nclist = nc;
%let selist = se1 se2 se3;
%let sclist = sc;
%mtfmax(ch10.ex101,ch10.ex101out);
proc format;
    value aa 1 = "Giancola & Zeichner, 1995"
             2 = "Gustafson, 1986        "
             3 = "Gustafson, 1991        "
             4 = "Lang et al., 1975      "
             5 = "Pihl & Zacchi, 1986    "
             6 = "Ratliff, 1984          "
             7 = "Zeichner & Pihl, 1979  "
             8 = "Zeichner & Pihl, 1980  ";
proc print data=ch10.ex101out noobs;
    title;
    format studyid aa. fmax pvalue 6.3;
run;
```

Output 10.1 F_{MAX} *statistics and p-values for individual studies in Table 10.2*

STUDYID	FMAX	NUMERDF	DENOMDF	PVALUE
Giancola & Zeichner, 1995	1.635	7	7	0.266
Gustafson, 1986	2.806	9	9	0.070
Gustafson, 1991	1.612	14	14	0.191
Lang et al., 1975	2.778	11	11	0.052
Pihl & Zacchi, 1986	1.855	15	15	0.121
Ratliff, 1984	1.272	5	5	0.399
Zeichner & Pihl, 1979	5.224	23	23	0.000
Zeichner & Pihl, 1980	2.641	23	23	0.012

As you can see in Output 10.1, the F_{MAX} statistic was significant for two of the eight studies. Thus, the equal variance assumption holds for most of the studies. To determine the impact of the unequal variances in the two studies by Zeichner and Pihl, we also use procedures based on unequal variances.

The SAS macro MULTRT(*indata,method*) in Appendix 10.2 was used to obtain an estimate and 95% confidence interval for each treatment effect size (that is, control versus placebo, control versus antiplacebo, control versus alcohol). *Indata* is the name of the input data file that contains the variable names for the means, standard deviations, and sample sizes for the control and experimental groups. Table 10.3 gives the variable names for this example.

Table 10.3 *Variable names for the means, standard deviations, and sample sizes for the control (C), placebo (E1), antiplacebo (E2), and alcohol (E3) groups*

Group	Mean	Standard deviation	Sample size
Control (C)	MC	SC	NC
Placebo (E1)	ME1	SE1	NE1
Antiplacebo (E2)	ME2	SE2	NE2
Alcohol (E3)	ME3	SE3	NE3

The METHOD parameter in the macro is coded 1 if the equal variance assumption is not met, or it is coded 2 if the equal variance assumption is met.

The following SAS code was used to print the results in Output 10.2 and Output 10.3. The %LET statements are similar to those described for the F_{MAX} statistic except there are two additional statements: (a) `%let melist = me1 me2 me3` gives the variable names for experimental group means, and (b) `%let mclist = mc` gives the variable name for control group mean. The statement `%multrt(ch10.ex101,2)` in line 8 specifies the name of the input data set, *indata* (that is, CH10.EX101), and the value of *method* (that is, 2 for equal variances).

```
libname ch10 "d:\metabook\ch10\dataset";
%let melist = me1 me2 me3;
%let nelist = ne1 ne2 ne3;
%let selist = se1 se2 se3;
%let mclist = mc;
%let nclist = nc;
%let sclist = sc;
%multrt(ch10.ex101,2);
proc format;
   value aa 1 = "QERROR";
   value bb 1 = "Placebo vs. control       "
            2 = "Antiplacebo vs. control "
            3 = "Alcohol vs. control       ";
proc print data=qerror noobs;
   title;
   format source aa.;
proc print data=bb noobs;
   format source bb.;
run;
```

Output 10.2 *Combined effect-size estimates and 95% confidence intervals for meta-analysis on alcohol-related aggression. Assumes equal control and experimental group variances.*

SOURCE	ESTIMATE	LOWER	UPPER
Placebo vs. control	0.12776	-0.13236	0.38789
Antiplacebo vs. control	0.43514	0.01676	0.85351
Alcohol vs. control	0.45401	0.12821	0.77981

As you can see in Output 10.2, the confidence intervals for the alcohol-versus-control and antiplacebo-versus-control comparisons do not include the value zero. Both groups that received alcohol were more physically aggressive than the control group. The confidence interval for the placebo-versus-control comparison included the value zero. Thus, alcohol-related expectancies did not increase physical aggression.

Output 10.3 *Heterogeneity statistic*

SOURCE	QSTAT	DF	PVALUE
QERROR	287.780	17	0

Because the *p*-value for the *Q*-statistic is less than .05, you reject the null hypothesis that the effects are homogeneous. However, you would not expect the effects to be homogeneous unless the three treatments do not differ.

For comparison purposes, we compute combined effect-size estimates and 95% confidence intervals using unequal variance procedures. The SAS code is the same as the code for Output 10.2 except that the value of *method* is set to 1 instead of 2. That is, the statement %multrt(ch10.ex101,1) is used in line 8. Output 10.4 shows the results.

Output 10.4 *Combined effect-size estimates and 95% confidence intervals for meta-analysis on alcohol-related aggression. Assumes unequal control and experimental group variances.*

SOURCE	ESTIMATE	LOWER	UPPER
Placebo vs. control	0.09806	-0.16655	0.36267
Antiplacebo vs. control	0.37382	-0.05668	0.80432
Alcohol vs. control	0.52770	0.20063	0.85477

Note that the confidence intervals that are based on unequal variances in Output 10.4 are wider than the confidence intervals that are based on equal variances in Output 10.2. Note also that the confidence interval for the antiplacebo-versus-control comparison includes the value zero in Output 10.4.

For comparison purposes, we also use meta-analytic procedures that assume independent effect sizes (see Chapter 8). Hedges' *g* is the effect-size estimate used. Output 10.5 shows the results from the following SAS code:

```
options nodate nocenter pagesize=54 linesize=80 pageno=1;
libname ch10 "d:\metabook\ch10\dataset";
data temp;
   set ch10.ex101;
   sp = sqrt(((nc-1)*sc*sc+(ne1-1)*se1*se1)/(nc+ne1-2));
   eff = (me1-mc)/sp;
   veff = (ne1+nc)/(ne1*nc)+(eff*eff)/(2*(ne1+nc-2));
   source = 1;
   output;
   sp = sqrt(((nc-1)*sc*sc+(ne2-1)*se2*se2)/(nc+ne2-2));
   eff = (me2-mc)/sp;
   veff = (ne2+nc)/(ne2*nc)+(eff*eff)/(2*(ne2+nc-2));
   source = 2;
   output;
   sp = sqrt(((nc-1)*sc*sc+(ne3-1)*se3*se3)/(nc+ne3-2));
   eff = (me3-mc)/sp;
   veff = (ne3+nc)/(ne3*nc)+(eff*eff)/(2*(ne3+nc-2));
   source = 3;
   output;
   keep source eff veff;
%let varlist = source;
%wavgeff(temp,tmpout,.95);
proc format;
   value aa 1 = "Placebo vs. control      "
            2 = "Antiplacebo vs. control "
            3 = "Alcohol vs. control      ";
proc print data=tmpout noobs;
   var source estimate level lower upper;
   format source aa. estimate level lower upper 6.3;
run;
```

Output 10.5 *Combined effect-size estimates and 95% confidence intervals for meta-analysis on alcohol-related aggression. Assumes independent effect-size estimates.*

SOURCE	ESTIMATE	LEVEL	LOWER	UPPER
Placebo vs. control	0.317	0.950	0.052	0.583
Antiplacebo vs. control	0.437	0.950	-0.035	0.910
Alcohol vs. control	0.773	0.950	0.442	1.104

As you can see in Output 10.5, the results are considerably different if you ignore the fact that the effect-size estimates are correlated. Both groups that expected alcohol were more aggressive than the control group. The confidence interval for the placebo-versus-control comparison included the value zero. Thus, the pure pharmacological effects of alcohol did not increase aggression. Note that the confidence intervals also are narrower in Output 10.2 than in Output 10.5.

10.3 Combining the Results from Multiple-Endpoint Studies

In multiple-endpoint studies there is a single experimental group (E) and a single control group (C), but p dependent variables or endpoints are measured for each participant in the study. You will rarely find a set of studies in which each study reports results for exactly the same set of endpoints. It is more likely that different studies will report results for different subsets of endpoints.

For each participant in the control group, the vector of endpoint measurements is assumed to have a multivariate normal distribution with mean vector

$\mu_C = \left(\mu_{C_1}, \dots, \mu_{C_p} \right)'$ and covariance matrix $\Sigma_C = \left\{ \sigma_{C_{jj'}} \right\}$. For each participant in the experimental group, the vector of endpoint measurements is assumed to have a multivariate normal distribution with mean vector $\mu_E = \left(\mu_{E_1}, \dots, \mu_{E_p} \right)'$ and covariance matrix $\Sigma_E = \left\{ \sigma_{E_{jj'}} \right\}$.

The effect size for the jth endpoint is

$$\delta_j = \frac{\mu_{E_j} - \mu_{C_j}}{\sqrt{\sigma_{C_{jj}}}}$$

(10.8)

(Gleser & Olkin, 1994, p. 347). The sample estimator of δ_j is

$$d_j = \frac{\bar{Y}_{E_j} - \bar{Y}_{C_j}}{\sqrt{S_{C_{jj}}}},$$

(10.9)

where $\bar{Y}_{E_j}$ and $\bar{Y}_{C_j}$ are the experimental and control group sample means for the jth endpoint, and $S_{C_{jj}}$ is the control group sample variance for the jth endpoint (Gleser & Olkin, 1994, p. 347). The effect-size estimates $d_1, \ldots, d_p$ are correlated because the endpoint measurements for any participant are correlated. The correlations among the d_j's depend on the correlations $\rho_{E_{jj*}}$ and $\rho_{C_{jj*}}$ between endpoints for participants in the experimental and control groups, respectively (Gleser & Olkin, 1994, p. 348). Because not all studies report sample correlations, the correlations must be imputed from other sources (for example, other studies, test manuals when the endpoints are subscores of a psychological test).

When the experimental and control group covariance matrices are homogeneous $\left(\text{that is,} \Sigma_E = \Sigma_C\right)$, a more precise effect-size estimator can be obtained by replacing $S_{C_{jj}}$ in Equation 10.9 with the pooled estimator

$$S_{POOLED_j} = \sqrt{\frac{(n_C-1)S_{C_{jj}} + (n_E-1)S_{E_{jj}}}{n_C - n_E - 2}}, \qquad (10.10)$$

where n_E and n_C the sample sizes for the experimental and control groups, respectively.

When the experimental and control group covariance matrices are homogeneous $\left(\text{that is,} \Sigma_E = \Sigma_C\right)$, the estimated large sample variance of d_j is

$$\text{Var}(d_j) = \frac{1}{n_E} + \frac{1}{n_C} + \frac{\frac{1}{2}d_j^2}{n_C + n_E}, \qquad (10.11)$$

and the estimated large sample covariance of d_j and d_{j*} $(j* \neq j)$ is

$$\text{Cov}(d_j, d_{j*}) = \left(\frac{1}{n_E} + \frac{1}{n_C}\right)r_{jj*} + \frac{\frac{1}{2}d_j d_{j*} r_{jj*}^2}{n_C + n_E}, \qquad (10.12)$$

where r_{jj*}^2 is the common value of the experimental and control correlations (Gleser & Olkin, 1994, p. 349).

When the experimental and control group variances are homogeneous but the covariances are not, the estimated large sample covariance of d_j and d_{j*} is

$$\text{Cov}\left(d_j, d_{j*}\right) = \frac{1}{n_E} r_{E_{jj*}} + \frac{r_{C_{jj*}}\left(1 + \frac{1}{2}\delta_j \delta_{j*} r_{C_{jj*}}\right)}{n_C}, \tag{10.13}$$

where $r_{E_{jj*}}$ and $r_{C_{jj*}}$ are the correlations in the experimental and control groups, respectively (Gleser & Olkin, 1994, p. 349).

If neither the variances nor covariances of the experimental and control groups are homogeneous, the large sample covariance of d_j and d_{j*} is

$$\text{Cov}\left(d_j, d_{j*}\right) = \frac{1}{n_E} \sqrt{\frac{S_{E_{jj}}}{S_{C_{jj}}}} \sqrt{\frac{S_{E_{j*j*}}}{S_{C_{j*j*}}}} r_{E_{jj*}} + \frac{r_{C_{jj*}}\left(1 + \frac{1}{2} d_j d_{j*} r_{C_{jj*}}\right)}{n_C}, \tag{10.14}$$

where $S_{E_{j*j*}}$ and $S_{C_{j*j*}}$ are the respective experimental and control group sample variances for endpoint $j*$, and $S_{E_{jj}}$ is the experimental group sample variance for endpoint j (Gleser & Olkin, 1994, p. 349).

***Example 10.2** Multiple endpoint studies*

To illustrate how to combine correlated effect-size estimates from multiple endpoint studies, we use a meta-analysis of the effects of coaching on Scholastic Aptitude Test (SAT) performance (Kalaian & Raudenbush, 1996). The SAT is a college entrance examination that contains verbal and math subtests called SAT-Verbal and SAT-Math, respectively. In this example, the treatment or experimental group received coaching on how to take the SAT, whereas the control group received no coaching. The data for the 47 studies are given in Table 10.4.

Table 10.4 *Studies on the effects of coaching on SAT performance*

Study	n_E	n_C	SAT-V g_U	SAT-M g_U
Alderman & Powers, 1980				
Evans & Pike, 1973				
Study 1	145	129	0.13	0.12
Study 2	72	129	0.25	0.06
Study 3	71	129	0.31	0.09
Laschewer, 1986	13	14	0.00	0.07
Roberts & Oppenheim, 1966				
Study 1	43	37	0.01	—
Study 2	19	13	0.67	—
Study 3	16	11	−0.38	—
Study 4	20	12	−0.24	—
Study 5	39	28	0.29	—
Study 6	38	25	—	0.26
Study 7	18	13	—	−0.41
Study 8	19	13	—	0.08
Study 9	37	22	—	0.39
Study 10	19	11	—	−0.53
Study 11	17	13	—	0.13
Study 12	20	12	—	0.26
Study 13	20	13	—	0.47
Zuman, 1988, Study 1	21	34	0.54	0.57
Zuman, 1988, Study 2	16	17	0.13	0.48
Burke, 1986				
Study 1	25	25	0.50	—
Study 2	25	25	0.74	—
Coffin, 1987	8	8	−0.23	0.33
Davis, 1985	22	21	0.13	0.13
Frankel, 1960	45	45	0.13	0.34
Kintisch, 1979	38	38	0.06	—
Whitla, 1962	52[a]	52[a]	0.09	−0.11
Curran, 1988				
Study 1	21	17	−0.10	−0.08
Study 2	24	17	−0.14	−0.29
Study 3	20	17	−0.16	−0.34
Study 4	20	17	−0.07	−0.06
Dear, 1958	60	526	−0.02	0.21

(continued)

Study	n_E	n_C	SAT-V g_U	SAT-M g_U
French, 1955				
Study 1	110	158	0.06	—
Study 2	161	158	—	0.20
FTC, 1978	192	684	0.15	0.03
Keefauver, 1976	16	25	0.17	−0.19
Lass, 1961	38	82	0.02	0.10
Reynolds & Oberman, 1987	93	47	−0.04	0.60
Teague, 1992	10	15	0.40	—

Note: SAT = Scholastic Aptitude Test; V = verbal; M = math. n_E = sample size for coached group. n_C = sample size for uncoached group. [a]The sample sizes for SAT-M were $n_E = n_C = 50$. FTC = Federal Trade Commission.

The SAS macro MULEND(*indata*) in Appendix 10.3 computes a combined effect-size estimate and 95% confidence intervals for each endpoint. The input data file, *indata*, contains the effect-size estimates for each endpoint, the control group sample sizes for each endpoint, and the experimental group sample sizes for each endpoint. The SAS code that is needed for computing the combined effect-size estimate and 95% confidence intervals for each endpoint (Output 10.6), and the Q statistic (Output 10.7) are given in this section.[1] Because Kalaian and Raudenbush (1996) used Hedges' unbiased estimator g_U (see Chapter 5, Section 5.2.1), which uses the pooled variance in the denominator, we assume that the equal variance assumption holds. The statement %let mlist = vgu mgu gives the effect-size estimate variable names for the two endpoints (that is, VGU = Hedges' g_U for SAT-Verbal, MGU = Hedges' g_U for SAT-Math). The statement %let nlist1 = vnc mnc gives the control group sample size variable names for the two endpoints (that is, VNC and MNC are the control group sample sizes for SAT-Verbal and SAT-Math, respectively). The statement %let nlist2 = vne mne gives the experimental group sample size variable names for the two endpoints (that is, VNE and MNE are the experimental group sample sizes for

---

[1] There are four more output data files that are not printed in this book: (a) the model matrix XXOUT, (b) the response vector DDOUT, (c) the model variance-covariance matrix VAROUT, and (d) the variance-covariance matrix for the effect-size estimates COVBB. These four files are not of interest to most users, but they can be printed.

SAT-Verbal and SAT-Math, respectively). The statement %LET CORR = 0.66 gives the correlation between the two endpoints.

```
libname ch10 "d:\metabook\ch10\dataset";
%let mlist = vgu mgu;
%let nlist1 = vnc mnc;
%let nlist2 = vne mne;
%let corr = 0.66;
%mulend(ch10.ex102);
proc format;
   value aa 1 = "QERROR";
   value bb 1 = "Verbal"
            2 = "Math  ";
proc print data=bb noobs;
   format source bb. estimate lower upper 6.3;
proc print data=qerror noobs;
   format source aa. qstat pvalue 6.3;
run;
```

Output 10.6 *Combined effect-size estimates and 95% confidence intervals for SAT coaching studies*

SOURCE	ESTIMATE	LOWER	UPPER
Verbal	0.123	0.061	0.184
Math	0.132	0.068	0.197

As you can see in Output 10.6, coaching produced a small-to-moderate increase in SAT-Verbal and SAT-Math scores. Neither confidence interval included the value zero.

Output 10.7 *Q-statistic for SAT coaching studies*

SOURCE	QSTAT	DF	PVALUE
QERROR	92.054	65	0.015

As you can see in Output 10.7, the *p*-value for the *Q*-statistic is less than .05, which suggests that the effects are not homogeneous.

For comparison purposes, we also use meta-analytic procedures that assume independent effect-sizes (see Chapter 8). Hedges' g_U was the effect-size estimate used. Output 10.8 shows the results.

Output 10.8 *Combined effect-size estimates and 95% confidence intervals for SAT coaching studies. Assumes independent effect-size estimates.*

SOURCE	ESTIMATE	LEVEL	LOWER	UPPER
Verbal	0.119	0.950	0.056	0.182
Math	0.129	0.950	0.060	0.197

In this example, the results in Output 10.8 are about the same as the results in Output 10.6. In general, however, it is important to consider the correlation between multiple endpoints when you calculate effect-size estimates.

10.4 Conclusions

In some studies, the effect-size estimates are not independent. Two common types of studies that produce correlated effect-size estimates are multiple-treatment and multiple-endpoint studies. Multiple-treatment studies compare more than one treatment with a common control. An effect-size estimate is calculated for each treatment versus control comparison. The effect-size estimates are correlated because of the common control group. In multiple-endpoint studies, there may be a single treatment group and a single control group, but multiple dependent variables are used as endpoints for each participant. A treatment versus control effect-size estimate is calculated for each dependent variable. Because endpoint measures on each participant are correlated, the effect-size estimates for the measures also are correlated within studies.

You should combine correlated effect-size estimates using multivariate procedures. Kalaian and Raudenbush (1996) describe three primary advantages of using a multivariate approach to combine correlated effect-size estimates from multiple-endpoint studies. First, a multivariate approach enables you to test whether an experimental treatment produces a larger effect for one dependent variable than for another. Second, a multivariate approach enables you to test whether study characteristics relate differently to different dependent variables. Third, a multivariate approach protects the meta-analyst against errors of inference that can occur when separate significance tests are performed for each dependent variable.

Similar advantages of multivariate procedures can be cited for multiple-treatment studies. First, a multivariate approach enables you to test which experimental treatment produces the largest effect. Second, a multivariate approach enables you to test whether study characteristics relate differently to different treatments. Third, a multivariate approach protects the meta-analyst against errors of inference that can occur when separate significance tests are performed for each treatment.

One disadvantage with using multivariate procedures is that they require knowledge of the correlations between dependent variables (in multiple-endpoint studies) and between treatments (in multiple-treatments studies). Unfortunately, these correlations are not always included in the research report. However, sometimes these correlations can be imputed. For example, in a multiple-endpoint study in which the endpoints are the subscales of a test, the correlations might be obtained from test manuals or from published literature on the test used.

10.5 References

Bushman, B. J. (1993). Human aggression while under the influence of alcohol and other drugs: An integrative research review. *Current Directions in Psychological Science, 2*, 148–152.

Bushman, B. J. (1997). Effects of alcohol on human aggression: Validity of proposed explanations. In D. Fuller, R. Dietrich, & E. Gottheil (Eds.), *Recent developments in alcoholism: Alcohol and violence* (Vol. 13, pp. 227–243). New York: Plenum.

Bushman, B. J. & Cooper, H. M. (1990). Effects of alcohol on human aggression: An integrative research review. *Psychological Bulletin, 107*, 341–354.

Buss, A. H., & Perry, M. (1992). The Aggression Questionnaire. *Journal of Personality and Social Psychology, 63*, 452–459.

Dollard, J., Doob, L. W., Miller, N. E., Mowrer, O. H., & Sears, R. R. (1939). *Frustration and aggression.* New Haven, CT: Yale University Press.

Giancola, P.R., & Zeichner, A. (1995). An investigation of gender differences in alcohol-related aggression. *Journal of Studies on Alcohol, 56*, 573–579.

Gleser, L. J., & Olkin, I. (1994). Stochastically dependent effect sizes. In H. Cooper & L. V. Hedges (Eds.). *The handbook of research synthesis* (pp. 339–355). New York: Russell Sage Foundation.

Graybill, F. A. (1976). Theory and Application of the Linear Model. Pacific Grove, California: Wadsworth & Brooks/Cole Advanced Books and Software.

Hartley, H. O. (1950). The maximum F-ratio as a short-cut test for heterogeneity of variance. *Biometrika, 37*, 308–312.

Kalaian, H. A., & Raudenbush, S. W. (1996). A multivariate mixed linear model for meta-analysis. *Psychological Methods, 1*, 227–235.

10.6 Appendices

Appendix 10.1: SAS Macro for Computing F_{MAX} Statistics in Multiple Treatment Studies

```
%macro mtfmax(indata,outdata);
/****************************************************************/
/*   INDATA: Input data file name                            */
/*                                                            */
/*   NELIST:  List of experimental group sample sizes        */
/*   SELIST:  List of experimental group standard            */
/*                 deviations                                 */
/*   NCLIST:  Control group sample size                      */
/*   SCLIST:  Control group standard deviation               */
/*                                                            */
/*   OUTDATA:  Output data file name                         */
/*     Output data file contains five variables:             */
/*     STUDYID:  Study identification number                 */
/*     FMAX:  Fmax statistic                                 */
/*     NUMERDF:  Numerator degrees of freedom for Fmax       */
/*            statistic                                       */
/*     DENOMDF:  Denominator degrees of freedom for Fmax     */
/*            statistic                                       */
/*     PVALUE:  p-value for Fmax statistic                   */
/****************************************************************/
proc iml;
   reset nolog;
   reset storage=ch10.imlrout;
   use &indata;
   read all var {&nelist} into nee;
   read all var {&nclist} into ncc;
   read all var {&selist} into see;
   read all var {&sclist} into scc;
   ncc = repeat(ncc,1,ncol(nee));
   scc = repeat(scc,1,ncol(see));
   nrowcc=nrow(ncc);
   studyid = (1:nrow(ncc))`;
/****************************************************************/
/*   Compute the F Test Statistics                           */
/****************************************************************/
   start compute;
   vee = see # see;
   vc = scc # scc;
   fvalue = vee / vc;
   df1 = nee-1;
   df2 = ncc-1;
   tmp=(nee = .);
   do i = 1 to ncol(fvalue);
      do j = 1 to nrow(fvalue);
         if ((fvalue[j,i] < 1) & (fvalue[j,i] ^= .)) then
         do;
            fvalue[j,i]=1/fvalue[j,i];
```

```
                df1[j,i]=ncc[j,i]-1;
                df2[j,i]=nee[j,i]-1;
            end;
            if (tmp[j,i]) then df2[j,i]=.;
        end;
    end;
/**********************************************************/
/*  Compute FMAX                                        */
/**********************************************************/
    fmax=j(nrow(fvalue),1,0);
    do j = 1 to nrow(fvalue);
        fmax[j]=max(fvalue[j,]);
    end;
    index=j(nrow(fvalue),1,0);
    do i = 1 to nrow(fvalue);
        do j = 1 to ncol(fvalue);
        if(fmax[i]=fvalue[i,j]) then
            do;
            index[i]=(i-1)*ncol(fvalue)+j;
            end;
        end;
    end;
    numerdf = df1[index];
    denomdf = df2[index];
    pvalue = 1 - probf(fmax,numerdf,denomdf);
/**********************************************************/
/*  Produce the Output                                  */
/**********************************************************/
    create &outdata var{studyid fmax numerdf denomdf pvalue};
    append;
    close &outdata;
    finish compute;
    run compute;
reset storage=ch10.imlrout;
store module=(compute);
run;
quit;
%mend mtfmax;
```

Appendix 10.2: SAS Macro for Computing Combined Effect-Size Estimates and 95% Confidence Intervals in Multiple Treatment Studies

```
%macro multrt(indata,method);
/**********************************************************/
/*    INDATA:   Input data set                           */
/*                                                        */
/*    NELIST:  List of experimental group sample sizes    */
/*    MELIST:  List of experimental group means           */
/*    SELIST:  List of experimental group standard        */
/*             deviations                                 */
/*    NCLIST: Control group sample size                   */
/*    MCLIST: Control group mean                          */
/*    SCLIST: Control group standard deviation            */
/*    METHOD:                                             */
/*      METHOD = 1: Control group variance                */
/*      METHOD = 2: Pooled variance                       */
/*                                                        */
/*  Output Data Files                                     */
/*    (1) DDOUT:  Response vector for the model            */
/*    (2) XXOUT:  Model matrix                            */
/*    (3) VAROUT: Model variance-covariance matrix         */
/*    (4) BB:  Effect-size estimate                        */
/*        SOURCE: SOURCE = p for the pth treatment        */
/*        EFFECT: Effect-size estimate for each           */
/*                treatment                               */
/*        LOWER: Lower bound of 95% confidence            */
/*               interval for each treatment              */
/*        UPPER: Upper bound of 95% confidence            */
/*               interval for each treatment              */
/*    (5) COVBBOUT:  Variance-covariance matrix for       */
/*                 each treatment                         */
/*    (6) QERROR:  Q statistic for the hypothesis          */
/*                 that all effect-sizes are equal         */
/*        SOURCE:   SOURCE = 1 for QERROR                  */
/*        QSTAT:  Q statistic                             */
/*        DF:    Degrees of freedom for Q statistic        */
/*        PVALUE:  p-value for Q statistic                 */
/**********************************************************/
proc iml;
    reset nolog;
    reset storage=ch10.imlrout;
    use &indata;
    read all var {&nelist} into nee;
    read all var {&melist} into mee;
    read all var {&selist} into see;
    read all var {&nclist &mclist &sclist};
    tmp1=j(nrow(mee),ncol(mee),0);
```

```
/*******************************************************/
/*  Compute the effect-size for each treatment and      */
/*   study                                              */
/*******************************************************/
   if (&method = 2) then do;
      sp = j(nrow(nee),1,0);
      np = j(nrow(nee),1,0);
      do i = 1 to nrow(nee);
       do j = 1 to ncol(nee);
        if (nee[i,j] ^= .) then do;
          np[i]=np[i]+nee[i,j]-1;
          sp[i]=sp[i]+see[i,j]*see[i,j]*(nee[i,j]-1);
         end;
        end;
        np[i] = nc[i]+np[i]-1;
        sp[i] = sp[i]+sc[i]*sc[i]*(nc[i]-1);
        sp[i] = sp[i]/np[i];
      end;
     end;

      do j = 1 to ncol(mee);
        if (&method = 1) then
           tmp1[,j] = (mee[,j]-mc) / sc;
        if (&method = 2) then
           tmp1[,j] = (mee[,j]-mc) / sp;
      end;
/*******************************************************/
/* Compute the estimated large-sample covariance for    */
/* each study                                          */
/*******************************************************/
    vdd=j(ncol(mee)*nrow(mee),ncol(mee),.);
    do k = 1 to nrow(mee);
       do i = 1 to ncol(mee);
          do j = 1 to ncol(mee);
           if (&method=1) then do;
             vdd[(k-1)*ncol(mee)+i,j] =
                (1+0.5*tmp1[k,i]*tmp1[k,j])/nc[k];
             end;
           else do;
             vdd[(k-1)*ncol(mee)+i,j] =
                1/nc[k]+(0.5*tmp1[k,i]*tmp1[k,j])/np[k];
             end;
             if (i = j) then vdd[(k-1)*ncol(mee)+i,j]
                = vdd[(k-1)*ncol(mee)+i,j] + 1/nee[k,i];
          end;
       end;
    end;
start datamod;
```

```
/**********************************************************/
/* Create the effect-size vector DD                      */
/**********************************************************/
    k=0;
    do i = 1 to ncol(tmp1);
        do j = 1 to nrow(tmp1);
            if (tmp1[j,i] = .) then k=k+1;
        end;
    end;
    var=j(nrow(vdd)-k,nrow(vdd)-k,0);
    index1=j(k,3,0);
    index2=j(nrow(vdd)-k,1,0);
    k1=1;
    k2=1;
    do I = 1 to nrow(tmp1);
        do J = 1 to ncol(tmp1);
            if (tmp1[i,j] = .) then
            do;
            index1[k1,1]=i;
            index1[k1,2]=j;
            index1[k1,3]=(i-1)*ncol(tmp1)+j;
            k1=k1+1;
            end;
        if (tmp1[i,j] ^= .) then
            do;
            index2[k2,1]=(i-1)*ncol(tmp1)+j;
            k2=k2+1;
            end;
        end;
    end;
    dd=remove(tmp1,index1[,3]);
    dd=dd`;
/**********************************************************/
/* Create the model matrix XX                            */
/**********************************************************/
    k=ncol(mee);
    indstep = j(1,k,1);
    do i = 1 to k;
        indstep[i]=i;
    end;
    xx=repeat(i(k),nrow(mee),1);
    xx=xx[index2 , indstep];
```

```
/*********************************************************/
/* Create the variance-covariance matrix VAR            */
/*********************************************************/
   rho=vdd[index2 , indstep];
   m=1;
   do I = 1 to nrow(rho);
      if (i = m) then
         do;
         k = I;
         index = I;
         do J = 1 to ncol(rho);
            if (rho[i,j] ^= .) then
               do;
               var[i,k] = rho[i,j];
               m=m+1;
               k=k+1;
               end;
            end;
         end;
      else
         do;
         k=index;
         do J = 1 to ncol(rho);
            if (rho[i,j] ^= .) then
               do;
               var[i,k] = rho[i,j];
               k=k+1;
               end;
            end;
         end;
      end;
/*********************************************************/
/* Produce the OUTPUT for the effect-size vector DD,    */
/* the model matrix XX, and the variance-covariance     */
/* matrix VAR                                           */
/*********************************************************/
   create ddout from dd;
   append from dd;
   close ddout;
   create xxout from xx;
   append from xx;
   close xxout;
   create varout from var;
   append from var;
   close var;
finish datamod;
reset noprint;
run datamod;
reset print;
start compute;
```

```
/*********************************************************/
/* Compute the generalized least square estimator:       */
/* (1) Effect-size estimates                             */
/* (2) Variance-covariance matrix of the effect-size     */
/*     estimates                                         */
/* (3) 95% confidence interval for effect-size           */
/*     estimate for each treatment                       */
/* (4) Test statistic for the hypothesis that all        */
/*     effect sizes are equal                            */
/*********************************************************/
   covbb=inv(xx`*inv(var)*xx);
   create covbbout from covbb;
   append from covbb;
   close covbbout;
   bb=covbb*xx`*inv(var)*dd;
   aa=i(ncol(xx));
   aaa=j(ncol(xx),1,1);
   lower=aa`*bb+probit(0.025)*diag(sqrt(aa`*covbb*aa))*aaa;
   upper=aa`*bb+probit(0.975)*diag(sqrt(aa`*covbb*aa))*aaa;
   estimate=bb;
   source = 1: ncol(see);
   create bb var{source estimate lower upper};
   append;
   close bb;
   qstat=dd`*inv(var)*dd-bb`*covbb*bb;
   df=nrow(xx)-ncol(xx);
   pvalue = 1 - probchi(qstat,df);
   Source = 1;
   create qerror var{source qstat df pvalue};
   append;
   close qstat;
   finish compute;
reset noprint;
run compute;
run;
quit;
%mend multrt;
```

Appendix 10: 3 SAS Macro for Computing Combined Effect-Size Estimates and 95% Confidence Intervals in Multiple End-Point Studies

```
%macro mulend(indata);
/*************************************************************/
/*   INDATA:   Input data set                            */
/*                                                       */
/*   MLIST:    List of effect size estimates for each end  */
/*             point                                     */
/*   NLIST1:   List of control group sample sizes for each */
/*             end point                                 */
/*   NLIST2:   List of experimental group sample sizes for */
/*             each end point                            */
/*   CORR: List of imputed correlationd between each     */
/*             pair of endpoints                         */
/*                                                       */
/*   Output Data Files                                   */
/*     (1) DDOUT:  Response vector for the model          */
/*     (2) XXOUT:  Model matrix                          */
/*     (3) VAROUT: Model variance-covariance matrix       */
/*     (4) BB:  Effect-size estimate                     */
/*         SOURCE: SOURCE = p for the pth end point       */
/*         EFFECT: Effect-size estimate for each          */
/*                 end point                             */
/*         LOWER: Lower bound of 95% confidence           */
/*                interval for each end point */           */
/*         UPPER: Upper bound of 95% confidence           */
/*                interval for each end point */           */
/*     (5) COVBBOUT:  Variance-covariance matrix for      */
/*                each end point */                       */
/*     (6) QERROR:  Q statistic for the hypothesis        */
/*                that all effect-sizes are equal         */
/*         SOURCE:   SOURCE = 1 for QERROR               */
/*         QSTAT:  Q statistic                           */
/*         DF:     Degrees of freedom for Q statistic     */
/*         PVALUE:  p-value for Q statistic              */
/*************************************************************/
proc iml;
   reset nolog;
   reset storage=ch10.imlrout;
   use &indata;
   read all var {&mlist} into effmean;
   read all var {&nlist1} into nendc;
   read all var {&nlist2} into nendt;
/*************************************************************/
/*  Compute the effect-size for each end point and        */
/*   study                                               */
/*************************************************************/
   tmp1=j(ncol(effmean)*nrow(effmean),1,.);
   do i = 1 to nrow(effmean);
      tmp1[2*i-1]=effmean[i,1];
      tmp1[2*i]=effmean[i,2];
   end;
```

```
/****************************************************/
/* Compute the estimated large-sample covariance for    */
/* each study under equal variance covariance assumption*/
/****************************************************/
   vdd=j(ncol(effmean)*nrow(effmean),ncol(effmean),.);
   do k = 1 to nrow(effmean);
      do i = 1 to ncol(effmean);
         do j = 1 to ncol(effmean);
            if (i = j) then
               do;
                 vdd[(k-1)*ncol(effmean)+i,j] = 1/nendc[k,i]
                 +1/nendt[k,i]+(0.5*effmean[k,i]
                 *effmean[k,i])/(nendc[k,i]+nendt[k,i]);
               end;
            else
               do;
                 vdd[(k-1)*ncol(effmean)+i,j] = (1/nendc[k,i]
                 +1/nendt[k,i])*&corr+(0.5*effmean[k,i]
                 *effmean[k,j]*&corr*&corr)
                 /(nendc[k,i]+nendt[k,i]);
               end;
         end;
      end;
   end;
start datamod;
/****************************************************/
/* Create the effect-size vector DD                      */
/****************************************************/
   k=0;
   do i = 1 to ncol(tmp1);
      do j = 1 to nrow(tmp1);
         if (tmp1[j,i] = .) then k=k+1;
      end;
   end;
   var=j(nrow(vdd)-k,nrow(vdd)-k,0);
   index1=j(k,3,0);
   index2=j(nrow(vdd)-k,1,0);
   k1=1;
   k2=1;
   do I = 1 to nrow(tmp1);
      do J = 1 to ncol(tmp1);
         if (tmp1[i,j] = .) then
         do;
         index1[k1,1]=i;
         index1[k1,2]=j;
         index1[k1,3]=(i-1)*ncol(tmp1)+j;
         k1=k1+1;
         end;
      end;
   end;
   dd=remove(tmp1,index1[,3]);
   dd=dd`;
```

```
/********************************************************/
/* Create the model matrix XX                          */
/********************************************************/
   k=ncol(effmean);
   xx=i(k);
   xx1=repeat(xx,nrow(effmean),1);
   k=ncol(xx1);
   do i = 1 to k;
     xx2=remove(xx1[,i],index1[,3]);
     if (i=1) then xx=xx2`;
     else xx = xx || (xx2`);
   end;
/********************************************************/
/* Create the variance-covariance matrix VAR          */
/********************************************************/
   k = ncol(vdd);
   do i = 1 to k;
     rho1=remove(vdd[,i],index1[,3]);
     if (i = 1) then rho=rho1`;
     else rho=rho||(rho1`);
   end;
   m=1;
   do I = 1 to nrow(rho);
      if (i = m) then
         do;
         k = I;
         index = I;
         do J = 1 to ncol(rho);
            if (rho[i,j] ^= .) then
               do;
               var[i,k] = rho[i,j];
               m=m+1;
               k=k+1;
               end;
            end;
         end;
      else
         do;
         k=index;
         do J = 1 to ncol(rho);
            if (rho[i,j] ^= .) then
               do;
               var[i,k] = rho[i,j];
               k=k+1;
               end;
            end;
         end;
      end;
   end;
```

```
/*************************************************************/
/* Produce the OUTPUT for the effect-size vector DD,       */
/* the model matrix XX, and the variance-covariance        */
/* matrix VAR                                              */
/*************************************************************/
    create ddout from dd;
    append from dd;
    close ddout;
    create xxout from xx;
    append from xx;
    close xxout;
    create varout from var;
    append from var;
    close var;
finish datamod;
reset noprint;
run datamod;
reset print;
start compute;
/*************************************************************/
/* Compute the generalized least square estimator:         */
/* (1) Effect-size estimates                               */
/* (2) Variance-covariance matrix of the effect-size       */
/*     estimates                                           */
/* (3) 95% confidence interval for effect-size             */
/*     estimate for each end point */
/* (4) Test statistic for the hypothesis that all          */
/*     effect sizes are equal                              */
/*************************************************************/
    covbb=inv(xx`*inv(var)*xx);
    create covbbout from covbb;
    append from covbb;
    close covbbout;
    bb=covbb*xx`*inv(var)*dd;
    aa=i(ncol(xx));
    aaa=j(ncol(xx),1,1);
    lower=aa`*bb+probit(0.025)*diag(sqrt(aa`*covbb*aa))*aaa;
    upper=aa`*bb+probit(0.975)*diag(sqrt(aa`*covbb*aa))*aaa;
    estimate=bb;
    source = 1: ncol(effmean);
    create bb var{source estimate lower upper};
    append;
    close bb;
    qstat=dd`*inv(var)*dd-bb`*covbb*bb;
    df=nrow(xx)-ncol(xx);
    pvalue = 1 - probchi(qstat,df);
    Source = 1;
    create qerror var{source qstat df pvalue};
    append;
    close qstat;
    finish compute;
reset noprint;
run compute;
run;
quit;
%mend mulend;
```

chapter 11

Conducting and Reporting the Results of a Meta-Analysis

11.1 Introduction

The results section of a meta-analysis should report (Halvorsen, 1994)

- the results of the literature search and the criteria that were used to select studies
- the results of the data collection, including descriptive statistics for various characteristics of the primary studies
- the results of the data analysis, including statistical assumption results, subgroup analysis results, and sensitivity analysis results.

In this chapter we first describe each type of result that is reported in a meta-analysis. Then we illustrate how to report meta-analytic results using a real example.

11.2 Reporting the Results of the Literature Search

The meta-analyst should explicitly state how the studies included in the meta-analysis were located. Without this information, the reader will not be able to determine whether the studies included in the meta-analysis are a representative sample of the population of studies that were conducted on the topic. When studies were located using computer searches, the meta-analyst should specify the databases that were searched, the years that were included in the searches, the keywords that were used to retrieve studies, and any restrictions placed on the search (for example, subject population=human, language=English). When studies were located using other methods, the meta-analyst also needs to describe the results from those methods.

Generally, all of the studies retrieved in a literature search are not included in a meta-analysis. Studies are sometimes excluded because they do not meet the definitions for the conceptual variables that are investigated in the meta-analysis, because they used poor methodologies, or because they do not contain enough information to calculate an effect-size estimate. We believe the last criteria is not a good reason for excluding studies if the studies do contain information about the direction and/or significance of results. If any studies were excluded, the meta-analyst is obligated to tell the reader what criteria were used to exclude studies. One way to present this information is in a table that lists the reasons for excluding studies from the meta-analysis and gives the number of studies excluded for each reason (Halvorsen, 1994). If you do not use a table to present exclusion/inclusion criteria, you should at least describe these criteria in the text of the manuscript. For example, in their meta-analysis on sex differences in leadership effectiveness, Eagly, Karau, and Makhijani (1995) used the following criteria for including studies.

> Criteria for including studies in the sample were that (a) the study compared male and female leaders, managers, supervisors, officers, department heads, or coaches on a measure of leadership effectiveness; (b) participants were at least 14 years old, from the United States or Canada, and not sampled from abnormal populations; (c) the study assessed the effectiveness of at least five leaders of each sex; (d) the sex of the leader and the sex of the subordinates were not completely confounded (as they would be in a study that compared female and male leaders of single-sex groups); and (e) the reported results were sufficient either to calculate a sex-of-participant effect size or to determine the statistical significance or direction of the sex difference. (p. 129)

You should list all of the studies included in the meta-analysis somewhere in the manuscript (usually in the reference section, marked with an asterisk), or you should tell readers where they can write to obtain a complete reference list.

11.3 Reporting the Results of the Data Collection

Meta-analysts should report the important information extracted from each study, including study characteristics that moderate effects, using a table. The most informative type of table reports each study in a separate row and each study characteristic in a separate column. For example, Table 11.1 lists a few rows for a meta-analysis on the effects of media violence on aggression in unconstrained social interaction (Wood, Wong, & Chachere, 1991, p. 376).

Table 11.1 Characteristics for each study in the meta-analysis on the effects of media violence on aggression in unconstrained social interaction

Study	Subject population	Setting[a]	Total N	Control conditions	Direction of effect	g_U[b]	95% CI Lower	95% CI Upper
Biblow (1973)	Fifth graders	1/1	40	Neutral film	$E>C$	0.32	-0.30	0.95
			40	No film	$C>E$	-0.47	-1.09	0.16
Cooper & Axsom (1981)	Kindergarteners to fifth graders	2/3	119	Neutral film				
Drabman & Thomas (1977)	Preschool boys	2/1	20	Prosocial film	$C>E$			
Ellis & Sekyra (1972)	First graders	2/2	51	Neutral film & no film combined	$E>C$	0.78	0.18	1.39
Fechter (1971)	Mentally retarded 8- to 38-year olds	2/4	40	Prosocial film	$E>C$			
Friedrich & Stein (1973)	Nursery schoolers	2/2	65	Neutral film	$E>C$			
			65	Prosocial film	$E>C$			

Note: Effect sizes are scored so that positive numbers reflect greater aggressiveness of subjects exposed to violent media, and negative numbers reflect greater aggressiveness of subjects in control conditions.

E = experimental condition; C = control.

[a]Setting is coded such that the first numeral reflects exposure setting (1 = experimentally constructed viewing room, 2 = classroom or viewing room in school, 3 = residence), and the second numeral reflects setting for behavioral observation of viewers' aggression (1 = experimentally constructed viewing room, 2 = classroom, viewing room in school, or school lunchroom, 3 = playground or athletic field, 4 = residence).

[b]Effect-size estimates are corrected to remove bias associated with small samples and are weighted by the inverse of the variance (Hedges & Olkin, 1985).

If the number of studies included in a meta-analysis is quite large, the editor might not allow the above table format because it requires too much journal space.

Another table format, one that requires less journal space but does not give as much information, reports frequency data for various study characteristics. For example, Table 11.2 lists study characteristic frequency data for a meta-analysis on sex differences in leadership effectiveness (Eagly et al., 1995, p. 132).

Table 11.2 Summary of study characteristics for meta-analysis on sex differences in leadership effectiveness

Variable and class	Value
Median publication year	1980
Publication form	
Journal article	52
Book or book chapter	5
Dissertation or master's thesis	28
Unpublished document	11
Median percentage of male authors	50
Median no. observations	112
Median no. judgments aggregated into each observation	8
Confounding of male-female comparison	
Controlled via matching	12
Known	27
Unknown and likely	34
Unknown and unlikely	23
Setting of study	
Laboratory	22
Organization	74
Business	22
Educational	21
Governmental or social service	7
Military	10
Miscellaneous	14
Size of group or organization	
Median no. members in laboratory group	3
Size of organization	
Small	13
Large	27
Mixed or unknown	34
Level of leadership	
First or line	43
Second or middle	22
Third or higher	1
Ambiguous, mixed, or unknown	30

(Continued)

Variable and class	Value
Median age of leaders	30
Laboratory	20
Organizational	38
Median percentage of men among leaders	60
Laboratory	50
Organizational	73
Median percentage of men among subordinates	48
Laboratory	50
Organizational	28
Basis of selection of laboratory leadership	
Appointed randomly	14
Emerged	5
Unclear or mixed	3
Basis of selection of organizational leaders	
Random sample	24
Unsuccessful random sample	26
Nonrandom	4
Unclear	20
Mean respondent judgments of roles	
Competence sex difference[a]	0.06
Interest sex difference	0.00
Stereotypic interest difference	0.11*
Interpersonal ability training[b]	10.56
Task ability rating	10.50

Note: For categorical variables, numbers in the table represent frequency of sex comparisons in each class. Summaries of continuous variables are based on reports for which information was available on each variable.

[a]For the first three variables constructed from judgments of leadership roles, values are positive for differences in the masculine direction (greater male estimates of competence and of interest; ascription of greater interest to average men).

[b]For the last two variables constructed from judgments of the leadership roles, values are larger to the extent that a role was judged to require more interpersonal or task ability (on 15-point scales with 15 indicating high ability).

* differs significantly ($p < .05$) from 0.00 (exactly no difference).

11.4 Reporting the Results of the Data Analysis

If the effect-size estimates in a meta-analysis are homogeneous, it is useful to report the mean effect-size estimate and a measure of its precision (for example, 95% confidence interval). When the meta-analysis includes a small number of studies, the effect-size estimate and 95% confidence interval for each study can be reported separately in either a table or in a dot plot (see Chapter 3, Section 3.2). We believe that a dot plot is more effective than a table because a dot enables readers to visually compare the location and width of confidence intervals for the individual studies. When the meta-analysis includes a large number of studies, the results can be presented using a stem-and-leaf plot or a box plot (see Chapter 3, Sections 3.5 and 3.6, respectively).

11.4.1 Checking Statistical Assumptions

The validity of meta-analytic procedures requires that certain assumptions be met. The normal quantile plot is useful for checking whether the standardized effect-size estimates come from a single, normal distribution (see Chapter 3, Section 3.4). If effect-size estimates are not independent, then you can use multivariate procedures to combine effect-size estimates (see Chapter 10). The reader of a meta-analysis is especially interested in knowing whether the variation in effect-size estimates across studies can be attributed to sampling alone or whether other sources (for example, study characteristics) contribute to the variation. Homogeneity tests enable you to determine whether the variation in effect-size estimates is larger than what you would expect by chance alone (see Chapters 8 and 9).

11.4.2 Reporting the Results of Subgroup Analyses

If the effect-size estimates are not homogeneous, then it is useful to divide effect-size estimates into subgroups. A format that works well for analyses that include results for subgroups of studies is shown in Table 11.3. Table 11.3 contains columns for source of heterogeneity, Q-test of homogeneity, degrees of freedom, mean effect-size estimate, standard error, and 95% confidence interval for subgroups of studies.

Table 11.3 *A table format for reporting results for subgroups of studies*

Source	Q	df	ES estimate	SE	95% CI
Between groups	$Q_{BETWEEN}$				
Within groups	Q_{WITHIN}				
Within group 1	Q_{WITHIN_1}				
Within group 2	$Q_{WITHIN2}$				
$\vdots$	$\vdots$				
Within group p	Q_{WITHIN_P}				
Total	Q_{TOTAL}				

Note: Q = homogeneity test statistic; df = degrees of freedom; ES = effect-size; SE = standard error; CI = confidence interval.

A table similar to Table 11.3 is given in the meta-analysis on sex differences in leadership effectiveness (Eagly et al., 1995, p. 136).

Table 11.4 *Categorical models predicting study-level effect sizes for meta-analysis on sex differences in leadership effectiveness*

Variable and class	Between-classes effect (Q_B)	k	d_+ and 95% CI	Homogeneity within each class (Q_{W_i})
Setting of study	1.81			
Laboratory		20	0.07 [-0.06, 0.20]	32.99*
Organization		56	-0.03 [-0.06, 0.01]	235.05*
Type of organization	92.92*			
Business		15	-0.07_a [-0.14, 0.00]	41.88*
Educational		16	-0.11_a [-0.18, -0.04]	61.29*
Governmental or social service		6	-0.15_a [-0.25, -0.05]	6.77
Military		7	0.42_b [0.32, 0.52]	15.58*
Miscellaneous[a]		10	-0.05_a [-0.16, 0.05]	9.84
Level of leadership	67.44*			
First or line		35	0.19_a [0.13, 0.26]	115.05*
Second or middle		20	-0.18_b [-0.24, -0.12]	42.11*
Ambiguous, mixed, or unknown		21	-0.03_c [-0.09, 0.02]	45.23*

Note: Effect-size estimates are positive for greater effectiveness of male leaders and negative for greater effectiveness of female leaders. Means with the same subscript are not significantly at the .05 level. k = number of studies; CI = confidence interval; d_+ = mean weighted effect-size estimate, where each estimate is weighted by the inverse of its variance.

[a]Includes hospitals and clinics, recreational camps, and mixed settings.

* $p < .05$.

As you can see in Table 11.4, although the setting of the study (that is, laboratory versus organization) did not moderate sex differences in leadership effectiveness, effects were heterogeneous for both settings. The type of organization influenced whether males or females were more effective leaders. Males were more effective leaders than females in the military, whereas females were more effective leaders than males in education and in government or social services. No sex differences in leadership effectiveness were observed in the other organizations. The effects were heterogeneous, however, for all types of organizations except government or social services and miscellaneous. Likewise, the level of leadership influenced whether males or females were more effective. Males were more effective than females in first or line leadership positions, whereas females were more effective than males in second or middle leadership positions. The effects for both levels of leadership, however, were heterogeneous. Because sex differences in leadership cannot be captured by a few simple study characteristics, a random-effects model might be more appropriate for these data than a fixed-effects model.

Parallel box plots are also quite useful in depicting the results from different subgroups of studies. Parallel box plots enable readers to see how the magnitude of effect-size estimates is related to systematic differences between studies.

11.4.3 Reporting the Results of Sensitivity Analyses

Readers want to know whether the results of a meta-analysis (for example, combined effect-size estimates, heterogeneity tests) are sensitive to the inclusion or exclusion of particular studies and whether the results are sensitive to the method of data analysis. Sensitivity analysis systematically addresses the question, "What happens if some aspect of the data or the analysis is changed?" (Greenhouse & Iyengar, 1994).

One question is whether studies that are potential outliers should be included in the combined analysis. Potential outliers can be identified using dot plots and box plots (see Chapter 3, Sections 3.2 and 3.6, respectively). In a dot plot, you can see whether the confidence interval for a given study overlaps with the confidence interval for all studies combined. You can use a box plot to detect mild outliers that are between 1.5 and 3 interquartile ranges from the box, and extreme outliers that are more than 3 interquartile ranges from the box.

To determine the influence of potential outlying effect-size estimates, you can report the combined results with and without these potential outliers. Sometimes, each study is removed from the meta-analysis to see if the results change as a function of that study's effect-size estimate (for example, Needleman & Gatsonis, 1990). For example, Eagly et al. (1995, p. 133) reported the results separately with potential outliers included in and excluded from the combined analysis (see Table 11.5). When potential outliers were included in the combined analysis, the effects were heterogeneous and the mean weighted effect-size estimate did not significantly differ from zero. When the 12 potential outliers were excluded from the combined analysis, the effects were homogeneous and the mean weighted effect-size estimate was significantly less than zero (indicating that females were more effective leaders than males). However, Eagly and her colleagues did not specify what constituted a potential outlier. It is generally a good practice to include potential outliers unless there is a very good reason to exclude them.

Table 11.5 *Combined analysis with and without outliers*

Criterion	Value
Known effect-size estimates	
k	76
Mean weighted d (d_+) and 95% CI	-0.02 [-0.05, 0.02]
Homogeneity (Q) of ds comprising d_+	269.85*
Mean unweighted d and 95% CI	-0.03 [-0.13, 0.07]
Median effect-size estimate	-0.07
Known effect-size estimates excluding outliers	
No. outliers removed	12
k	64
Mean weighted d (d_+) and 95% CI	-0.12 [-0.16, -0.08]
Homogeneity (Q) of ds comprising d_+	86.30

Note: Effect-size estimates are positive for greater effectiveness of male leaders and negative for greater effectiveness of female leaders. k = number of studies; CI = confidence interval; d = effect-size estimate; d_+ = mean weighted effect-size estimate, where each estimate is weighted by the inverse of its variance; Q = homogeneity of effect estimates.

* $p < .05$.

Stem-and-leaf plots and box-plots are also useful in detecting skewness in the distribution of effect-size estimates (see Chapter 3, Sections 3.5 and 3.6, respectively). If the data are severely skewed, a transformation might be necessary (Tukey, 1977). In a sensitivity analysis, you can determine whether the combined results differ as a function of the data transformation.

You can use the funnel plot and normal quantile plot to detect publication bias (see Chapter 3, Sections 3.3 and 3.4, respectively). Once publication bias has been detected, you should determine whether the results of the meta-analysis are sensitive to the publication bias mechanism (see Greenhouse & Iyengar, 1994).

You must also make several decisions about analysis procedures, such as whether to use a fixed- or random-effects model and whether to use a weighted or unweighted mean to estimate the population effect size. You can test whether the results of the meta-analysis are sensitive to such decisions. We recommend that meta-analysts fit both fixed- and random-effects models to the same data set. If the variation in effect-size estimates can be captured by a few simple study characteristics, a fixed-effects model is preferred because it is more powerful than a random-effects model (Rosenthal, 1995). We recommend the use of a weighted mean to estimate the population effect size because studies with smaller sample sizes should produce less accurate effect-size estimates than studies with larger sample sizes.

11.5 Example of a Meta-Analysis

Informal observation suggests that some people are especially likely to become involved in aggressive interactions. The personality trait of aggression is referred to as trait aggressiveness. In this example, taken from a meta-analysis by Anderson and Bushman (1997), we examine whether trait aggressiveness is related to actual aggressive behavior.

Trait aggressiveness can be defined operationally using (a) self-report personality scales, (b) aggression nominations by others (for example, peers, teachers, counselors), and (c) violent histories. The most widely used self-report measure of trait aggressiveness is the Buss-Durkee Hostility Inventory (BDHI; Buss & Durkee, 1957). Sample items from the BDHI include "Once in a while I cannot control my urge to harm others" and "Whoever insults me or my family is asking for a fight."

The relation between trait aggressiveness and aggressive behavior has been investigated both outside and inside of the laboratory. Outside the laboratory, scores on self-report measures of aggressiveness are compared for people with and without violent histories (for example, criminals, patients, adolescent offenders). Inside the laboratory, scores on self-report measures of aggressiveness are correlated with the level of noxious stimuli (for example, electric shocks, noise blasts, heat pulses). Because laboratory researchers study more homogeneous populations (for example, college students), trait aggressiveness should vary more outside of the laboratory. In other words, trait aggressiveness has a restricted range in laboratory studies. Thus, stronger relations between trait aggressiveness and aggressive behavior might be expected in field studies than in laboratory studies.

11.5.1 Reporting the Results of the Literature Search

The PsycINFO computer database was searched (1974–1997) using the keywords aggress* and violen*. The asterisk at the end of the keyword gives all forms of the keywords (for example, aggressive, aggression). The aggress* and violen* keywords were paired with the keywords Buss, BDHI, Aggression Questionnaire, AQ, and trait aggress*. The search was restricted to studies published in English and to studies that used human participants.

The literature search resulted in 29 studies that examined the relation between trait aggressiveness and aggressive behavior. All 29 studies reported enough information to compute correlation coefficients. Of the 29 studies, 15 were conducted in laboratory settings and 14 were conducted in field settings.

11.5.2 Reporting the Results of the Data Collection

Table 11.6 lists the correlations for the 29 studies on trait aggressiveness and aggressive behavior. The table contains the reference, sample size, correlation, and setting for each study.

Table 11.6 *Studies correlating trait aggressiveness with aggressive behavior*

Study	N	r	Context
Shemberg et al. (1968)	45	.59	1
Knott (1970)	18	.51	1
Williams et al. (1967)	60	.43	1
Giancola & Zeichner (1995a)	79	.42	1
Wolfe & Baron (1971)	40	.35	1
Scheier et al. (1978)	163	.34	1
Hammock & Richardson (1992)	196	.28	1
Giancola & Zeichner (1995b)	60	.27	1
Hartman (1969)	57	.25	1
Bushman (1995)	296	.23	1
Leibowitz (1968)	38	.23	1
Pihl et al. (1997)	114	.20	1
Cleare & Bond (1995)	48	.15	1
Larsen et al. (1972)	78	.15	1
Muntaner et al. (1990)	85	−.02	1
Boone & Flint (1988)	53	.61	2
Lange et al. (1995)	49	.52	2
Maiuro et al. (1988)	67	.50	2
Selby (1984)	100	.46	2
Buss & Perry (1992)	98	.45	2
Gunn & Gristwood (1975)	30	.45	2
Maiuro et al. (1988)	68	.44	2
Boone & Flint (1988)	52	.44	2
Maiuro et al. (1988)	58	.41	2
Lothstein & Jones (1978)	61	.40	2
Renson et al. (1978)	51	.39	2
Stanford et al. (1995)	214	.38	2
Archer et al. (1995)	100	.33	2
Syverson & Romney (1985)	60	.27	2

Note: N = sample size; r = correlation coefficient; Context: 1 = laboratory study, 2 = field study.

11.5.3 Reporting the Results of the Data Analysis

First, a dot plot is constructed for the 29 studies that are listed in Table 11.6, rank ordered by the magnitude of the correlation coefficient separately for laboratory and field settings. A dot plot effectively displays the estimate and 95% confidence interval for the population correlation coefficient for each study in the meta-analysis. The following SAS code produces the dot plot in Output 11.1:

```
options nodate nocenter pagesize=54 linesize=132 pageno=1;
libname ch11 "d:\metabook\ch11\dataset";
data;
    set ch11.ex1;
    lower = eff + probit(0.025)*sqrt(veff);
    upper = eff + probit(0.975)*sqrt(veff);
    lower = (exp(2*lower)-1)/(exp(2*lower)+1);
    upper = (exp(2*upper)-1)/(exp(2*upper)+1);
proc sort;
    by context eff;
proc format;
    value aa 1 = "Laboratory"
             2 = "Field       ";
proc timeplot;
    plot lower="[" rr="*" upper="]"/
        overlay hiloc ref=0 refchar="0";
    id study context;
    format context aa.;
    title;
run;
```

Output 11.1 *Dot plot for studies correlating trait aggressiveness with aggressive behavior*

As you can see from Output 11.1, all but one of the correlations are positive, and the correlations appear to be larger for field studies than for laboratory studies. None of the field studies have 95% confidence intervals that include the value zero, whereas five out of 15 laboratory studies have 95% confidence intervals that include the value zero.

Next, a homogeneity test was conducted to determine whether the sample correlations for the 29 studies in the meta-analysis were homogeneous. Study context (that is, laboratory versus field) was ignored in this analysis. The following SAS code was used to conduct the homogeneity test. Output 11.2 shows the results.

```
options nodate nocenter pagesize=54 linesize=132 pageno=1;
libname ch11 "d:\metabook\ch11\dataset";
data;
    set ch11.ex1;
    weight = 1/ veff;
    dummy = 1;
proc glm noprint outstat=ch11.ex1b;
    class dummy;
    model eff = dummy / noint;
    weight weight;
data ch11.ex1b;
    source = "ERROR";
    set ch11.ex1b;
    if (_type_ = "ERROR");
    qstat = ss;
    pvalue = 1 - probchi(qstat,df);
    keep source df qstat pvalue;
proc print noobs;
    title;
    format qstat pvalue 6.3;
run;
```

Output 11.2 *Test of homogeneity of effects for studies correlating trait aggressiveness with aggressive behavior*

SOURCE	DF	QSTAT	PVALUE
ERROR	28	50.017	0.006

As you can see in Output 11.2, the Q-statistic is significant at the .05 level, indicating that the 29 sample correlations are not homogeneous. Note that you can also obtain the Q-statistic in this analysis by using the WEIGHT option in PROC GLM. The Q-statistic is given as the corrected total sum of squares. The advantage of the previous SAS code is that it only prints out the Q-statistic and it also calculates a p-value for the Q-statistic. The SAS GLM procedure does not provide a p-value associated with the corrected total sums of squares. For comparison purposes, PROC GLM was used to conduct the homogeneity test. Output 11.3 shows the ANOVA table portion of the results from the following SAS code:

```
options nodate nocenter pagesize=54 linesize=80 pageno=1;
libname ch11 "d:\metabook\ch11\dataset";
data;
   set ch11.ex1;
   weight = 1 /veff;
proc glm;
   class context;
   model eff = context;
   weight weight;
quit;
```

Output 11.3 *ANOVA table portion output from the GLM procedure in SAS software*

General Linear Models Procedure
Dependent Variable: EFF
Weight: WEIGHT

Source	DF	Sum of Squares	Mean Square	F Value	Pr > F
Model	1	18.34021428	18.34021428	15.63	0.0005
Error	27	31.67711560	1.17322650		
Corrected Total	28	50.01732988			

As you can see in Output 11.3, the corrected total sum of squares is the same value as the Q-statistic that is given in Output 11.2.

11.5.3.1 Checking Statistical Assumptions

First, construct a normal quantile plot to check the normality assumption (see Chapter 3, Section 3.4.1). Because the distribution of the correlation coefficient is not normal when $\rho \neq 0$, Fisher's (1921) z transformation was first applied to the correlations. Each Fisher's z score was divided by its standard deviation (that is, $1/\sqrt{n-3}$ to obtain a standardized effect-size estimate.

The following SAS code produces the normal quantile plot shown in Figure 11.1. To produce Figure 11.1 with the SAS macro CIQQPLOT, discussed in Section 3.4 of Chapter 3, you need to specify the range of the standardized effect-size estimates in the AXIS1 statement. The following SAS code finds the range of the standardized effect-size estimates:

```
options nodate nocenter pagesize=54 linesize=80 pageno=1;
libname ch11 "d:\metabook\ch11\dataset";
data;
    set ch11.ex1;
    stdeff = eff / sqrt(veff);
proc univariate noprint;
    var stdeff;
    output out=outa min=mineff max=maxeff;
proc print data=outa noobs;
    title;
    var mineff maxeff;;
    format maxeff mineff 7.4;
quit;
```

Output 11.4 *Range of standardized effect-size estimates for meta-analysis on the relation between trait aggressiveness and aggressive behavior*

MINEFF	MAXEFF
-0.1811	5.8112

As you can see in Output 11.4, the largest value of the standardized effect size was about 6 and the smallest value was slightly less than 0. Thus, the ORDER option was set at −2 to 7 in the AXIS1 statement. The following SAS code produces Figure 11.1:

```
libname ch11 "d:\metabook\ch11\dataset";
libname gdevice0 "c:\tmp";
filename fig1101 "c:\tmp\fig1101.cgm";
goptions reset=goptions device=cgmmwwc gsfname=fig1101
    gsfmode=replace ftext=hwcgm005 gunit=pct;
data ch11.fig1101;
   set ch11.ex1;
   stdeff = eff / sqrt(veff);
%ciqqplot(ch11.fig1101,c11graph,fig1101,stdeff);
```

Figure 11.1 *Normal quantile plot for studies correlating trait aggressiveness with aggressive behavior*

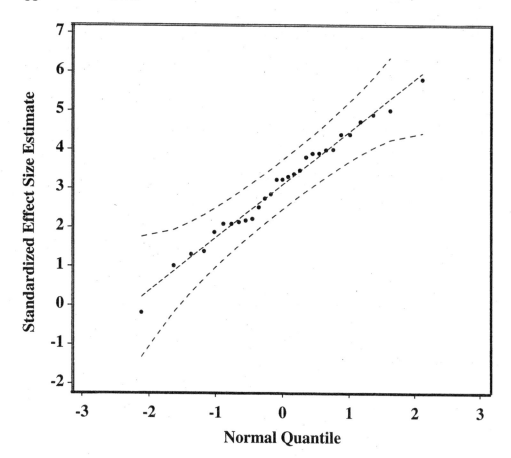

As you can see in Figure 11.1, the standardized effect-size estimates appear to fall on a straight line and within the 95% confidence limits. However, half of the standardized effect-size estimates are greater than 3, suggesting that they probably do not come from the same normal population. In a single normal population, it is very unlikely that half of the effects would exceed 3 standard deviation units.

The following SAS code constructs a stem-and-leaf plot. A stem-and-leaf plot is very useful in examining the shape of the distribution of effects (see Chapter 3, Section 3.5). Output 11.5 displays the stem-and-leaf portion of the output that is generated by PROC UNIVARIATE.

```
options nodate nocenter pagesize=54 linesize=80 pageno=1;
libname ch11 "d:\metabook\ch11\dataset";
proc univariate  data=ch11.ex1 plot;
   var eff;
run;
```

Output 11.5 *Stem-and-leaf plot for studies correlating trait aggressiveness with aggressive behavior*

```
Univariate Procedure
Variable=EFF
    Stem Leaf
       7 1
       6 8
       5 0568
       4 0124567788
       3 457
       2 0336889
       1 55
       0
      -0 2
         ----+----+----+----+
         Multiply Stem.Leaf by 10**-1
```

The distribution of effect-size estimates that is displayed in Output 11.5 appears to be bimodal. One peak occurs at the stem 4, whereas the other peak occurs at the stem 2. Thus, it appears that the effects do not come from a single normal population.

Based on the dot plot in Output 11.1, the normal quantile plot in Figure 11.1, and the stem-and-leaf plot in Output 11.5, we strongly believe that the context of the study should be considered rather than ignored in the analyses. That is, we believe that the effects should be divided into two subgroups: laboratory studies and field studies.

11.5.3.2 Reporting the Results of Subgroup Analyses

Because the data in Table 11.6 are not homogeneous, a moderator analysis was performed to determine whether study context explained some of the excess variation in sample correlations. The following SAS code calculates the Q-statistics for the ANOVA table (see Chapter 8, Section 8.4.1.1 and Appendix 8.1). Output 11.6 shows the results.

```
libname ch11 "d:\metabook\ch11\dataset";
%let varlist1 = context;
%let varlist2 = context;
%within(ch11.ex1,ch11.out1106,eff,veff,context,2);
proc format;
    value aa  1 = "Between Groups      "
              2 = "Within Groups       "
              3 = "  Within (Laboratory)"
              4 = "  Within (Field)     "
              5 = "Corrected Total      ";
proc print noobs;
   title;
   format source aa. qstat pvalue 7.3;
quit;
```

Output 11.6 ANOVA table for heterogeneity summary statistics for studies correlating trait aggressiveness with aggressive behavior

SOURCE	DF	QSTAT	PVALUE
Between Groups	1	18.340	0.000
Within Groups	27	31.677	0.244
Within (Laboratory)	14	23.071	0.059
Within (Field)	13	8.606	0.802
Corrected Total	28	50.017	0.006

The significant between-groups Q statistic in Output 11.6 indicates that correlations are larger for field studies than for laboratory studies. In addition, both types of studies were homogeneous (that is, both within-group tests were non-significant). Note also that the Q-statistic for the corrected total is equal to the Q-statistic for error in Output 11.2.

Next, we estimate the population correlation coefficient and 95% confidence interval for each type of study context. Output 11.7 shows the results from the following SAS code (see Chapter 8, Section 8.4.1.2 and Appendix 8.2):

```
options nodate nocenter pagesize=54 linesize=80 pageno=1;
libname ch11 "d:\metabook\ch11\dataset";
data temp;
   set ch11.ex1;
   output;
   context = 3;
   output;
%let varlist = context;
%wavgeff(temp,ch11.out1107,.95);
data;
   set ch11.out1107;
   estimate=(exp(2*estimate)-1)/(exp(2*estimate)+1);
   lower=(exp(2*lower)-1)/(exp(2*lower)+1);
   upper=(exp(2*upper)-1)/(exp(2*upper)+1);
   rename fact1=context;
proc format;
   value aa 1 = "Laboratory Study"
            2 = "Field Study      "
            3 = "Over All         ";
proc print noobs;
   var context estimate level lower upper;
   format context aa. estimate level lower upper 6.3;
run;
```

Output 11.7 *Confidence intervals for studies correlating trait aggressiveness with aggressive behavior inside and outside of the laboratory*

CONTEXT	ESTIMATE	LEVEL	LOWER	UPPER
Laboratory Study	0.264	0.950	0.211	0.315
Field Study	0.423	0.950	0.371	0.472
Over All	0.338	0.950	0.301	0.374

As you can see in Output 11.7, trait aggressiveness is positively correlated with aggression in both study contexts, but the relation is stronger outside of the laboratory than inside of the laboratory. This difference in study contexts is probably due to the fact that the range of trait aggressiveness scores in laboratory studies is restricted because participants are generally college students. Thus, it appears that the effects do not come from a single normal population.

In the stem-and-leaf plot that is depicted in Output 11.5, the distribution of correlations is bimodal with one peak at stem 4 and the other peak at stem 2. These peaks correspond with the mean correlations of .423 and .264 for field and laboratory studies, respectively.

Because study context moderated the relation between trait aggressiveness and aggressive behavior, it is useful to depict the results graphically using parallel box plots. The following SAS code produces the side-by-side box plots in Figure 11.2 (see Chapter 3, Section 3.6):

```
/*  side-by-side box plots;                        */
options nodate pagesize=54 linesize=80 nocenter pageno=1;
libname ch11 "d:\research\metabook\ch11\dataset";
data fig1102;
   set ch11.ex1;
   label eff="Effect Size Estimate";
   label context="  ";
proc sort data=fig1102;
   by context;
filename fig1102 "d:\tmp\fig1102.cgm";
goptions reset=goptions device=cgmmwwc gsfname=fig1102
   gsfmode=replace ftext=hwcgm005 gunit=pct;
proc shewhart data=fig1102 gout=fig1102 graphics;
   boxchart eff*context / boxstyle=schematic
                          idsymbol=dot
                          vref=0
                          haxis=axis1
                          nolegend
                          hoffset=5
                          nolimits
                          stddevs
                          boxwidth=30;
axis1 value =   ("Laboratory Studies" " " "Field Studies");
run;
quit;
```

Figure 11.2 *Parallel box plots for field and laboratory studies correlating trait aggressiveness with aggressive behavior*

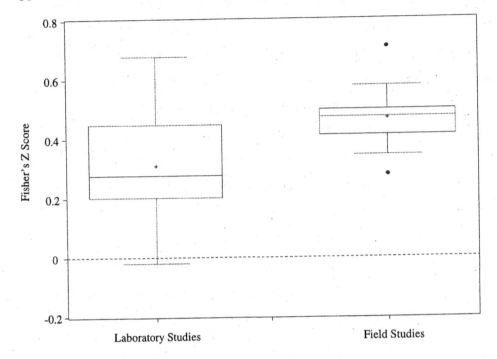

As you can see in Figure 11.2, the relation between trait aggressiveness and aggressive behavior is stronger in field studies than in laboratory studies. The distributions for both study contexts are skewed to the right, and there are two mild outlying effects for field studies.

11.5.3.3 Reporting the Results of Sensitivity Analyses

A sensitivity analysis was performed to determine if the pooled results were affected by the potential outliers. From the dot plot in Output 11.1, it appears that there are two potential outlying laboratory studies (Muntaner et. al., 1990; Shemberg et. al., 1968). The 95% confidence intervals for both of these studies do not overlap with the 95% confidence interval for all studies combined (see Output 11.7). In the parallel box plots in Figure 11.2, there are two potential outliers for field studies: one positive (Boone & Flint, 1988) and one negative (Syverson & Romney, 1985).

Output 11.8 contains the ANOVA table for heterogeneity summary statistics with these four potential outliers removed. The pattern of results in Output 11.8 is the same as the pattern of results in Output 11.3. The between-groups Q statistic is significant, indicating that effect-size estimates were larger for field studies than for laboratory studies. In addition, both types of studies were homogeneous (that is, both within-group tests were nonsignificant).

Output 11.8 *ANOVA table for heterogeneity summary statistics for studies correlating trait aggressiveness with aggressive behavior with four potential outliers removed*

SOURCE	DF	QSTAT	PVALUE
Between Groups	1	14.819	0.000
Within Groups	23	12.698	0.958
Within (Laboratory)	12	9.152	0.690
Within (Field)	11	3.545	0.981
Corrected Total	24	27.517	0.281

Output 11.9 contains the confidence intervals for laboratory studies, field studies, and all studies combined with the four potential outliers deleted. The pattern of results is the same as in Output 11.7. The average effect-size estimates are about the same, the width of the confidence intervals is about the same, and none of the confidence intervals contain the value zero.

Output 11.9 *Confidence intervals for studies correlating trait aggressiveness with aggressive behavior with four potential outliers removed*

CONTEXT	ESTIMATE	LEVEL	LOWER	UPPER
Laboratory Study	0.269	0.950	0.214	0.323
Field Study	0.420	0.950	0.366	0.472
Over All	0.340	0.950	0.301	0.378

Removing the four potential outliers does not change the pattern of results. Thus, the results of the meta-analysis are not sensitive to the influence of the four potential outliers.

We also attempted to determine whether the study results depend on whether a fixed- or random-effects model is used. The computer program did not converge, however, because the random-effects variance was very close to zero. Because the random-effects variance is about zero, a fixed-effects analysis is appropriate for this data set.

11.5.4 Conclusions about Example Meta-Analysis

Overall, trait aggressiveness was significantly correlated with aggressive behavior. The correlation was stronger, however, for field studies than for laboratory studies, perhaps because the range of trait aggressiveness scores is restricted in laboratory studies. A fixed-effects model appears to be appropriate because the variation in effect-size estimates can be captured by a single study characteristic (that is, laboratory versus field setting) and because the estimate of the random-effects variance is about zero. The same pattern of results was obtained when the four mild potential outliers were removed from the data set. We recommend that you include potential outliers in the meta-analysis unless you have a very good reason for excluding them.

11.6 Conclusions

The heart of a meta-analysis is the results section. A results section should report the results of the literature search and the criteria used to select studies. Without this information, it is impossible for the reader to determine whether the studies included in the meta-analysis are a representative sample of the population of studies conducted on the topic.

The results section of a meta-analysis also should report the results of the data collection. When possible, this should be accomplished using a table that reports each study in a separate row and each study characteristic in a separate column. The table also should report the effect-size estimate and 95% confidence interval for each study. Better yet, you can use a dot plot to display the effect-size estimates and 95% confidence intervals for each study. An alternative table format, one that requires less journal space but does not give as much information, reports frequency data for various study characteristics.

The results section of a meta-analysis also should report the results of the data analysis, including the results from statistical assumption tests and plots, subgroup analyses, and sensitivity analyses. You can use both descriptive and inferential statistics to report the results of the data analysis. You can use the normal quantile plot and heterogeneity test to assess whether the studies come from a single population. The normal quantile plot also can test whether the distribution of effect-size estimates is normally distributed. You can use parallel box plots and moderator tests to assess the relation between study characteristics and effect-size estimates. The meta-analyst should also test whether the results of a meta-analysis are sensitive to the inclusion or exclusion of particular studies (for example, potential outliers) and whether the results are sensitive to the method of data analysis (for example, fixed- versus random-effects model).

This chapter also illustrated how to report meta-analytic results using a real example. We hope this example helps clarify how to perform a meta-analysis. We wish you luck on your own data sets. Happy "meta-analyzing!"

11.7 References

Anderson. C. A., & Bushman, B. J. (1997). External validity of "trivial" experiments: The case of laboratory aggression. *Review of General Psychology, 1,* 19–41.

Boone, S. L., & Flint, C. (1988). A psychometric analysis of aggression and conflict-resolution behavior in black adolescent males. *Social Behavior and Personality, 16,* 215–226.

Buss, A. H., & Durkee, A. (1957). An inventory for assessing different kinds of hostility. *Journal of Consulting Psychology, 21,* 343–349.

Eagly, A. H., Karau, S. J., & Makhijani, M. G. (1995). Gender and the effectiveness of leaders: A meta-analysis. *Psychological Bulletin, 117,* 125–145.

Fisher, R. A. (1921). On the 'probable error' of a coefficient of correlation deduced from a small sample. *Metron, 1,* 1–32.

Greenhouse, J. B., & Iyengar, S. (1994). Sensitivity analysis and diagnoistics. In H. Cooper & L. V. Hedges (Eds.). *The handbook of research synthesis* (pp. 383–409). New York: Russell Sage Foundation.

Halvorsen, K. T. (1994). The reporting format. In H. Cooper & L. V. Hedges (Eds.). *The handbook of research synthesis* (pp. 425–437). New York: Russell Sage Foundation.

Hedges, L. V., & Olkin, I. (1985). *Statistical methods for meta-analysis.* New York: Academic Press.

Muntaner, C., Walter, D., Nagoshi, C., Fishbein, D., Haertzen, C. A., & Jaffe, J. H. (1990). Self-report vs. laboratory measures of aggression as predictors of substance abuse. *Drug and Alcohol Dependence, 25,* 1–11.

Needleman, H. L., & Gatsonis, C. A. (1990). Low-level lead exposure and the IQ of children. Journal of the American Medical Association, 263, 673–678.

Raudenbush, S. W. (1984). Magnitude of teacher expectancy effects on pupil IQ as a function of the credibility of expectancy induction: A synthesis of findings from 18 experiments. *Journal of Educational Psychology, 76,* 85–97.

Rosenthal, R. (1995). Writing meta-analytic reviews. *Psychological Bulletin, 118,* 183–192.

Shemberg, K. M., Leventhal, D. B., & Allman, L. (1968). Aggression machine performance and rated aggression. *Journal of Experimental Research in Personality, 3,* 117–119.

Syverson, K. L., & Romney, D. M. (1985). A further attempt to differentiate violent from nonviolent offenders by means of a battery of psychological tests. *Canadian Journal of Behavioural Science, 17,* 87–92.

Tukey, J. W. (1977). *Exploratory data analysis.* Reading, MA: Addison-Wesley.

Wood, W., Wong, F. Y., & Chachere, J. G. (1991). Effects of media violence on viewers' aggression in unconstrained social interaction. *Psychological Bulletin, 109,* 371–383.

Index

Books Available from SAS Press

Advanced Log-Linear Models Using SAS®
by **Daniel Zelterman**

Analysis of Clinical Trials Using SAS®: A Practical Guide
by **Alex Dmitrienko, Geert Molenberghs, Walter Offen,** *and*
Christy Chuang-Stein

Analyzing Receiver Operating Characteristic Curves with SAS®
by **Mithat Gönen**

Annotate: Simply the Basics
by **Art Carpenter**

Applied Multivariate Statistics with SAS® Software,
Second Edition
by **Ravindra Khattree**
and **Dayanand N. Naik**

Applied Statistics and the SAS® Programming Language,
Fifth Edition
by **Ronald P. Cody**
and **Jeffrey K. Smith**

An Array of Challenges — Test Your SAS® Skills
by **Robert Virgile**

Building Web Applications with SAS/IntrNet®: A Guide to the
Application Dispatcher
by **Don Henderson**

Carpenter's Complete Guide to the SAS® Macro Language,
Second Edition
by **Art Carpenter**

Carpenter's Complete Guide to the SAS® REPORT Procedure
by **Art Carpenter**

The Cartoon Guide to Statistics
by **Larry Gonick**
and **Woollcott Smith**

Categorical Data Analysis Using the SAS® System,
Second Edition
by **Maura E. Stokes, Charles S. Davis,**
and **Gary G. Koch**

Cody's Data Cleaning Techniques Using SAS® Software
by **Ron Cody**

Common Statistical Methods for Clinical Research with
SAS® Examples, Second Edition
by **Glenn A. Walker**

The Complete Guide to SAS® Indexes
by **Michael A. Raithel**

CRM Segmentation and Clustering Using SAS® Enterprise
Miner™
by **Randall S. Collica**

Data Management and Reporting Made Easy with
SAS® Learning Edition 2.0
by **Sunil K. Gupta**

Data Preparation for Analytics Using SAS®
by **Gerhard Svolba**

Debugging SAS® Programs: A Handbook of Tools and
Techniques
by **Michele M. Burlew**

Decision Trees for Business Intelligence and Data Mining: Using
SAS® Enterprise Miner™
by **Barry de Ville**

Efficiency: Improving the Performance of Your SAS®
Applications
by **Robert Virgile**

The Essential Guide to SAS® Dates and Times
by **Derek P. Morgan**

The Essential PROC SQL Handbook for SAS® Users
by **Katherine Prairie**

Fixed Effects Regression Methods for Longitudinal Data
Using SAS®
by **Paul D. Allison**

Genetic Analysis of Complex Traits Using SAS®
Edited by **Arnold M. Saxton**

A Handbook of Statistical Analyses Using SAS®, Second Edition
by **B.S. Everitt**
and **G. Der**

Health Care Data and SAS®
by **Marge Scerbo, Craig Dickstein,**
and **Alan Wilson**

The How-To Book for SAS/GRAPH® Software
by **Thomas Miron**

In the Know ... SAS® Tips and Techniques From
Around the Globe, Second Edition
by **Phil Mason**

Instant ODS: Style Templates for the Output Delivery System
by **Bernadette Johnson**

Integrating Results through Meta-Analytic Review Using
SAS® Software
by **Morgan C. Wang**
and **Brad J. Bushman**

Introduction to Data Mining Using SAS® Enterprise Miner™
by **Patricia B. Cerrito**

Learning SAS® by Example: A Programmer's Guide
by **Ron Cody**

support.sas.com/pubs

SAS® Macro Programming Made Easy, Second Edition
by **Michele M. Burlew**

SAS® Programming by Example
by **Ron Cody**
and **Ray Pass**

SAS® Programming for Researchers and Social Scientists,
Second Edition
by **Paul E. Spector**

SAS® Programming in the Pharmaceutical Industry
by **Jack Shostak**

SAS® Survival Analysis Techniques for Medical Research,
Second Edition
by **Alan B. Cantor**

SAS® System for Elementary Statistical Analysis,
Second Edition
by **Sandra D. Schlotzhauer**
and **Ramon C. Littell**

SAS® System for Regression, Third Edition
by **Rudolf J. Freund**
and **Ramon C. Littell**

SAS® System for Statistical Graphics, First Edition
by **Michael Friendly**

The SAS® Workbook and Solutions Set
(books in this set also sold separately)
by **Ron Cody**

Saving Time and Money Using SAS®
by **Philip R. Holland**

Selecting Statistical Techniques for Social Science Data:
A Guide for SAS® Users
by **Frank M. Andrews, Laura Klem, Patrick M. O'Malley,
Willard L. Rodgers, Kathleen B. Welch,**
and **Terrence N. Davidson**

Statistical Quality Control Using the SAS® System
by **Dennis W. King**

Statistics Using SAS® Enterprise Guide®
by **James B. Davis**

A Step-by-Step Approach to Using the SAS® System
for Factor Analysis and Structural Equation Modeling
by **Larry Hatcher**

A Step-by-Step Approach to Using SAS® for Univariate and
Multivariate Statistics, Second Edition
by **Norm O'Rourke, Larry Hatcher,**
and **Edward J. Stepanski**

Step-by-Step Basic Statistics Using SAS®: Student Guide
and Exercises
(books in this set also sold separately)
by **Larry Hatcher**

Survival Analysis Using SAS®: A Practical Guide
by **Paul D. Allison**

Tuning SAS® Applications in the OS/390 and z/OS
Environments, Second Edition
by **Michael A. Raithel**

Univariate and Multivariate General Linear Models:
Theory and Applications Using SAS® Software
by **Neil H. Timm**
and **Tammy A. Mieczkowski**

Using SAS® in Financial Research
by **Ekkehart Boehmer, John Paul Broussard,**
and **Juha-Pekka Kallunki**

Using the SAS® Windowing Environment: A Quick Tutorial
by **Larry Hatcher**

Visualizing Categorical Data
by **Michael Friendly**

Web Development with SAS® by Example, Second Edition
by **Frederick E. Pratter**

Your Guide to Survey Research Using the SAS® System
by **Archer Gravely**

JMP® Books

Elementary Statistics Using JMP®
by **Sandra D. Schlotzhauer**

JMP® for Basic Univariate and Multivariate Statistics: A Step-by-
Step Guide
by **Ann Lehman, Norm O'Rourke, Larry Hatcher,**
and **Edward J. Stepanski**

JMP® Start Statistics, Third Edition
by **John Sall, Ann Lehman,**
and **Lee Creighton**

Regression Using JMP®
by **Rudolf J. Freund, Ramon C. Littell,**
and **Lee Creighton**

Example Code — Examples from This Book at Your Fingertips

You can access the example programs for this book by linking to its companion Web site at **support.sas.com/companionsites**. Select the book title to display its companion Web site, and select **Example Code and Data** to display the SAS programs that are included in the book.

For an alphabetical listing of all books for which example code is available, see **support.sas.com/bookcode**. Select a title to display the book's example code.

If you are unable to access the code through the Web site, send e-mail to **saspress@sas.com**.

Comments or Questions?

If you have comments or questions about this book, you may contact the author through SAS as follows.

Mail: SAS Institute Inc.
SAS Press
Attn: <Author's name>
SAS Campus Drive
Cary, NC 27513

E-mail: saspress@sas.com

Fax: (919) 677-4444

Please include the title of the book in your correspondence.

See the last pages of this book for a complete list of books available through **SAS Press** or visit **support.sas.com/pubs.**

SAS Publishing News: Receive up-to-date information about all new SAS publications via e-mail by subscribing to the SAS Publishing News listserv. Visit **support.sas.com/subscribe.**

CPSIA information can be obtained
at www.ICGtesting.com
Printed in the USA
LVOW04s1943110717
540919LV00002B/12/P